Cryogenic Recycling and Processing

Editor

Norman R. Braton

Assistant Chairman and Professor
Department of Mechanical Engineering
University of Wisconsin
Madison, Wisconsin

CRC Press, Inc.
Boca Raton, Florida

Library of Congress Cataloging in Publication Data

Main entry under title:

Cryogenic recycling and processing.

Bibliography: p.
Includes Index.
1. Low temperature engineering. 2. Manufactoring processes. 3. Recycling (Waste, etc.) I. Braton, Norman R.
TP482.C76 621.5'9 80-13290
ISBN 0-8493-5779-9

Direct all inquiries to CRC Press, Inc., 2000 N.W. 24th Street, Boca Raton, Florida, 33431.

International Standard Book Number 0-8493-5779-9

Library of Congress Card Number 80-13290
Printed in the United States

PREFACE

The purpose of this book is to promote useful knowledge in the field of cryogenics. To accomplish this, the manuscript presents a composite of the state-of-the-art knowledge, research, and application of cryogenic processes.

Experts in the field were invited to contribute chapters. Appreciation goes to James A. Koutsky, Robert Kisielewski, David R. Otis, and James M. Venetucci for giving so much of their time in the preparation of their individual chapters. Their dedication to the goal of this publication made possible the meeting of all deadlines which is so necessary for completing a manuscript.

Typical of all manuscripts of its kind, *Cryogenic Recycling and Processing* owes its existence to more than its contributing authors and I would be remiss in my duties if I did not mention some key individuals.

Certainly much credit and appreciation must be given to Dean W. Robert Marshall, College of Engineering, University of Wisconsin—Madison. His foresight in recognizing the potential for cryogenic recycling and his subsequent full-hearted support of the cryogenic research program at the University of Wisconsin—Madison have substantially contributed to the content of this book.

Others at University of Wisconsin—Madison who contributed in important ways to completion of the book are Professor Edward F. Obert, Mechanical Engineering Department, whose encouragement and council aided in planning the contents of the book; Lynda L. Parker, Assistant Director, Space Science and Engineering Center, whose editorial expertise gave the text form and consistency; and Jan Larsen whose typing transformed our tattered rough drafts into a finished manuscript.

Lastly, a most heartfelt thanks is extended to the CRC staff — B. J. Starkoff, Benita Budd Segraves, Donna Grosso, and Sandy Pearlman — whose expertise and advice have been extremely helpful during the manuscript preparation.

FOREWORD

CRYOGENIC RECYCLING AND PROCESSING TODAY AND TOMORROW

Within one decade, the 1970s, cryogenic processing of solid wastes was conceived, born, and matured. By 1979 every major country had researched and experimented with this relatively new approach to solving age-old problems in the area of recycling. Four major U.S. industrial gas companies, namely Air Products, Airco, Liquid Carbonic, and Union Carbide, are prepared to supply total-system technology. Advances in improved system efficiencies, noise reductions, and automation are important breakthroughs in interesting greater numbers of industries to invest in cryogenic processing.

Perhaps no one idea in the history of the University of Wisconsin—Madison caught on so quickly and attracted as much national and international attention as did the concept for recycling of difficult items by cryogenics. Since its conception 10 years ago, representatives from every major country have come to the University of Wisconsin—Madison to learn more about cryogenic recycling, approximately 50 papers on the subject have been written with more than 2000 copies distributed, and approximately fifty students and nine faculty members at the University of Wisconsin—Madison have been directly involved in some phase of cryogenic research.

The advent of cryogenics for recycling has opened many avenues for processing what was considered to be difficult to recycle items, such as the automobile and truck tire. The tire, as we know it, has been manufactured in its present form since 1894. During the late 1960s and early 1970s in the U.S. alone, an estimated 234,000,000 tires were being discarded annually. This is approximately one for every man, woman, and child in the country. Each disposed tire occupies 3 ft^3 of space on the average. With landfill sites available at premium prices, this becomes quite costly. Tires are not biologically degradable, therefore, do not disintegrate, and when they are placed in sanitary landfills they do not stay buried. Because of their "breathing-type" action, tires will eventually resurface when passed over by compactors.

However, even if the above problems were to be overcome, tires are made from valuable materials and should not be abandoned. When manufactured, each tire requires approximately 8 gal of oil, 5 go into the manufacture of the tire and 3 are used as energy in the curing. The rubber from cyrogenically recycled tires can be used in the manufacturing of a number of products; surfacing for roads, and surfacing for flat roofs and runway surfaces, to name just a few. The use of the rubber from just one used tire should prolong the life of an average-sized driveway by several hundred percent. Of course, tires are only one of many items which can be more easily and economically processed cyrogenically than by other techniques.

Almost weekly, new concepts for uses of cryogenic technology are made known. In the area of recycling, items made from one or more materials which become embrittled at reduced temperatures have the potential for cryogenic processing and separation. The use of cryogenic processing in manufacturing is maturing rapidly, however, its potential in this field is not yet fully realized. The use of cold temperatures in foundries for the precision removal of sprues and risers has received widespread acceptance. The method of reducing temperatures for shrink type fits has been used for decades. The use of cryogens to keep the cutting edges of tools for material removal is being researched. This kind of technology, when working with metals, can extend tool life manyfold, when working polymers, the cold tool will prevent gumming.

Medical technology is finding cold temperatures to have advantages in many applications. The use of cryogens for surgery of the eye is well known, whereas the use of reduced temperatures for the storage of organs and limbs is more recent. The cold storage of semen for cattle breeding purposes has been most successful, in fact so successful that improved quality from crossbreeding has yielded offspring that exceed the quality of the original prize bull. Therefore, in some rare instances, the need for and importance of cryogenic technology has already exceeded its usefulness. Then, of course, the topic of cryogenic cemeteries is good for a few laughs at any cocktail party. Whereas the science for reversible storage of biomaterials has a promising future, the freezing of bodies for later revival doesn't hold much hope. It is suggested that one not bet one's life on it.

The freezing of foods for storage and future use was first patented by Mr. Birdseye in the mid-1930s. Fifty years later it is big business and still growing. As common a practice as food freezing is, new developments are continually surfacing, such as the use of fluidized bedding of salts for even more rapid freezing of whole sides of beef. In bridge and dam construction, it is often a problem to know what to do with the water and muck while new footings and foundations are being poured. Contractors now solve this problem simply by freezing that portion adjacent to the material to be removed. The list of cryogenic technological developments is long and growing. Within the time it will take to get this book off the press the list of other uses for cryogens, unthought of today, will perhaps once again double.

Cryogenic technology is today still in its infancy, perhaps at the stage the furnaces were 1000 years ago.

THE EDITOR

Norman R. Braton is Director of the Cryogenic Research Laboratory, Assistant Chairman, and Coordinator of the Materials-Joining Courses in the Department of Mechanical Engineering at the University of Wisconsin—Madison.

Professor Braton was a Captain, Ordnance, U.S. Army, from 1941 to 1946. He graduated in 1946 from the University of Minnesota—St. Cloud, with a B.A. degree, and from the University of Minnesota—St. Paul in 1953 with an M.A. degree in Industrial Education. He joined the University of Wisconsin faculty in 1955, and has since been active in the teaching and research and development of materials-joining and recycling.

Professor Braton became involved in cryogenic recycling in the early seventies, and established the first cryogenic laboratory devoted to recycling materials. In the next 10 years, Professor Braton lectured on the subject to more than 140 audiences throughout the world. In 1976, he organized the first international conference of its kind on the subject of cryogenic recycling. In 1979, a summary of his work was presented on video tape at the annual American Society of Mechanical Engineers meeting in New York and received 1st Place in the Theme Film Festival.

Professor Braton is a member of the American Welding Society and the American Society for Engineering Education. He has authored several texts, numerous articles, and holds four patents. He is a registered Professional Engineer with the State of Wisconsin.

Among his awards are the Faculty Service Award of the National University Extension Association (1979), the American Society of Mechanical Engineers Forum Theme Film Festival Award (1979), the Adams Memorial Award (1962), the Charles H. Jennings Memorial Award (1968), and the Twelfth District Meritorious Award (1966) of the American Welding Society.

CONTRIBUTORS

Norman R. Braton
Assistant Chairman and Professor
Department of Mechanical Engineering
University of Wisconsin
Madison, Wisconsin

Robert Kisielewski
Assistant Professor
Division of Technology
Arizona State University
Tempe, Arizona

James A. Koutsky
Professor
Department of Chemical Engineering
University of Wisconsin
Madison, Wisconsin

David R. Otis
Professor
Department of Mechanical Engineering
University of Wisconsin
Madison, Wisconsin

Jim M. Venetucci
Manager, Technical Sales Department
Industrial Gas Division
Liquid Carbonic Corporation
Chicago, Illinois

TABLE OF CONTENTS

Chapter 1

RECYCLING OF SOLID WASTES: A HISTORY

N. Braton

TABLE OF CONTENTS

I. INTRODUCTION

The crowned dragon that eats its own tail (or World Snake)* is an ancient alchemical symbol of self-renewal and regeneration.[1] Since the earliest days of ironmaking, dating back to 3000 B.C., man has been cognizant of the value of metal to society. Metal's physical properties allowed it to be shaped readily into tools, utensils, and weapons.[2] When damaged metal objects could not be repaired, the metal could be reclaimed by remelting. This regeneration of metal and other products has come to be known as recycling — the transformation of waste materials into usable products. This chapter deals with the history of recycling of solid wastes in the U.S., and it most logically begins with the recycling of metals.

Intertwined with the growing importance of metal in our civilization is the growth of the industry that enabled metal makers to reuse the surplus material left over in the manufacturing process and reclaimed from obsolete objects.

II. EVOLUTION OF SCRAP METAL INDUSTRY IN AMERICA

A. The Junk Collectors

Scrap use came to North America in 1642, when the first iron furnace on the continent was built in Saugus, Mass. Although the growth of the ironmaking industry in the colonies was slow and not without considerable hardship, the eastern coast of the new country was soon dotted with small iron works that relied mainly on swamp or bog ore.

The products of these early foundries included cast iron forms and crudely rolled plates. The plates were sheared in strips or rods to be converted by hand into nails. One grade of early scrap, "nail whiskers," derived its name from this product.

The value of scrap was understood by the early iron manufacturers. There is evidence that iron purchasers in colonial times were required to return the scrap they generated in their fabricating process to the iron works they dealt with.

By the 1820s, the growing demand for metal products had already produced a surplus of products which were worn out or discarded. Peddlers or junk collectors would make forays out into the rural areas to barter for worn-out farm implements and other items and resell the metal to mills and foundries for remelting.

In the mid-1800s, the sale of scrap was still limited because of steel mill technology. At that time the Bessemer process, which used very little scrap in its melts, was dominant. A huge reservoir of scrap started to build up, and many dealers and processors found no buyers for their products.

A revolution in the steelmaking industry took place in the latter part of the 19th century as the open hearth furnace gradually supplanted the Bessemer process. Some experts on the history of steelmaking feel that the development of a large reservoir of scrap spurred the growth of the open hearth process. The advent of the open hearth, and later, the electric furnace resulted in the start of a dramatic rise in scrap demand. The modern scrap industry was born, and the men who were to pioneer processing methods had set the course.

The history and origins of the scrap business are closely tied to those who entered the business during that transitional period. Scrap was an unglamorous, demanding, low-profit endeavor, requiring persistence and hard work. Few men wished to invest

* The illustration was adapted from an alchemical text: Abraham Eleazar (Abraham the Jew), *Urlates Chymisches Werk,* Leipzig, 1760, as reproduced in Jung, C. G., *Psychology and Alchemy,* Pantheon Books, New York, 1953, 99. With permission.

their lives in this risky business. The scrap industry might never have developed except that during this period of time, a large number of immigrants seeking refuge from the tyranny of their homelands came to the U.S. Most arrived with little or no money and had little formal training in recognized skills. To them fell the hard task — but a longed-for opportunity — of making a livelihood with limited or no resources.

Many of these newcomers to the U.S. learned the scrap business by spending years as collectors and peddlers themselves. After they had saved a sufficient amount of money and gained the necessary knowledge, some ventured into the risky business of being city dealers.

The city dealer in the late 1800s and the early 1900s was the central point for scrap processing. Purchasing scrap from all available sources including collectors, factories, and country dealers, the task of sorting and preparation fell to this new businessman.

Hiring unskilled labor, sometimes on a contract basis, the city dealer separated his acquired scrap into five classes: malleable, machinery cast, stove plate, melting steel, and rolling mill scrap.

Early city dealers had little, if any, equipment to help in the task of preparing scrap. Chisels and sledge hammers were used to break apart the larger metal objects. Even the force of freezing water was used by early scrap processors, who would pour water into the cracks of cast pieces and wait for the freezing water to break the metal apart.

As the demand for scrap grew, city dealers increased their productive capacity by introducing innovative new methods of handling scrap. Some dealers used explosives to break up large iron castings. Another method to break up large pieces into charge box size was the use of a ''skull cracker''. These were early cranes used to raise a heavy iron ball which was dropped from a considerable height onto the heavy castings. A gold medal at the 1904 St. Louis World's Fair was awarded for the first successful scrap-handling magnet, and within 2 years the magnets were used throughout the industry.

Considered for many years to be one of the greatest innovations in the industry, the gas cutting torch was first used in a scrap yard in Lebanon, Pa. about 1910. It allowed dealers to cut scrap without large expensive equipment.

B. Modern Scrap Industry

The birth of the modern scrap industry took place when collectors stopped selling directly to the mills and foundries themselves and took their wares to centrally located depots, where a new type of businessman was hard at work preparing metal to be used more efficiently by those same mills and foundries.

In the early days, there was not a clear division between collector and processor.

Many of the large scrap companies of today can trace their ancestry back to the horse-drawn cart days. Some of the processors had been exclusively collectors before they realized that by preparing scrap to make it more easily used by mills, they could increase the value of their products. Prior to this, many mills and foundries had processed scrap themselves. It was not until the early 1900s that nearly all mills and foundries abandoned this operation in favor of purchasing prepared scrap. It was also during this transition that collectors and dealers started buying from metal fabricators for processing and resale.

As dealers and processors grew in number, and mills bought from more than one seller, standardization was introduced in order to insure that when a contract was made, both parties were talking about the same commodity. The first attempt to separate the different types of scrap was to label the material collected from peddlers, factories, and farms as "country mixed" and the metal from the railroads as "railroad mixed".

Although mechanical improvements in scrap processing continued to be important in the modernization of the industry, eventually leading up to huge capital investments, the real essence of the growth of the industry was the persistence and determination of the men who made the transition from peddlers and collectors into large-scale scrap processing businesses.

Although the adversities were many, the foresighted dealer realized that the use of scrap metal was growing. While no one could predict that it would reach the size it has in the 20th century, many could see the trend.

The scrap-utilization picture started to change again when in 1910 the electric furnace, still in its earliest infancy, was recognized as an important factor in the future of steelmaking. The demand for alloy steel created by the growing automotive industry prompted interest in electric furnace production that was to have a profound effect on the scrap industry many years later. Unlike the Bessemer and open hearth processes, the electric furnace was not dependent on pig iron for its ferrous charge; it could use virtually 100% scrap.

The old methods of scrap handling were proved to be inadequate, and a transition to the modern era was undertaken. Many companies in the scrap industry set their future course during this period. With the cost of the new equipment extremely high, even by the standards of today, scrap processors were forecasting that the new demand levels would be sustained long enough to recover the large capital outlay. To some, the high demand period lasted long enough to recoup their investment; others lost severely as the economy fell into the depression of the 1930s.

Companies that survived that era saw steel demand rise again in the late 1930s, as prosperity returned to the country and preparation began for a war that appeared imminent.

Another wave of equipment purchases and improvements came with the large demand for scrap immediately before World War II; but like earlier booms, the increased demand for scrap was not sustained. In the early 1950s, yet another change in steelmaking methods forced the scrap market down sharply. Mills were introducing the basic oxygen furnace, which unlike the open hearth or electric furnaces, used only a small percentage of scrap. The scrap it did use was primarily home scrap — scrap generated in the mill during the manufacture of new steel.

Demand was stimulated again during the Korean War, but the pattern of a rising and falling marketplace for iron and steel scrap had become clear. In the 1960s, regeneration of metal products through the scrap metal industry was on the decline.[3] This was due largely to the following factors:

1. The cost of virgin raw materials to the manufacturer is almost as low as the cost of recycled materials, and virgin materials are usually qualitatively superior to salvage. Consequently, demand for secondary materials is limited.
2. Natural resources are abundant and manufacturing industries have deployed their operations and perfected their technologies to exploit these. No corresponding deployments and technology to exploit wastes have developed.
3. Natural resources occur in concentrations while wastes occur in a dispersed manner. Consequently, acquisition of wastes for recycling is costly.
4. Virgin materials, even in unprocessed form, tend to be more homogeneous in composition than waste materials. Sorting of wastes is costly and, in an age of affluence and convenience, repugnant to those who would have to engage in it — the urban householders.
5. The advent of synthetic materials made from hydrocarbons, and their combination with natural materials, cause contamination of the latter, limiting their recovery. The synthetics themselves at this time are virtually impossible to sort and recover economically.

In addition to the above factors affecting recycling, syndicated columnist Sylvia Porter points out there are also political deterrents affecting recycling:[4]

1. Licensing and zoning laws prohibit recyclers from maintaining operations close to their urban waste sources and the industrial manufacturers they supply.
2. There are little or no tax incentives similar to those given other industries to help promote investment in recycling equipment.
3. There are few or no government policies to provide guidelines which would boost to a maximum the content of recycled materials in all products and supplies purchased by state and municipal agencies.
4. Intrastate transportation rates on recycled materials are clearly discriminatory, subjecting recycled materials to charges that are higher than those on competing commodities.
5. Finally, and most importantly, most states and municipalities fail to recognize the vital role that the technical, operating, and marketing knowledge of the recycling industry can play in assuring the long-term success of resource recovery programs.

C. Environmental Impact of Scrap Metal Industry

Scrap contains the energy and iron necessary to make new iron and steel products.[5]

When 908 metric tons (1000 tons) of steel are made with scrap, the energy saved is equivalent to 529,900 ℓ (140,000 gal) of gasoline. By extending this figure to the record 46.5804 million metric tons (51.3 million tons) of scrap purchased during 1974 for use in the U.S., we find that the energy savings realized from using that scrap iron was equivalent to nearly 27.252 billion ℓ (7.2 billion gal) of gasoline — enough to power 20 million automobiles (48.27 km to 3.785 ℓ [30 mi to the gallon] nearly 17,699 km (11,000 mi) each. (This represents nearly 20% of the passenger automobiles registered in the U.S. in 1974.) Average number of miles driven by an automobile is about 16,000 km (10,000 mi)/year.

Each 0.908 metric ton (1 ton) of scrap used in making new iron and steel also saves 1.362 metric tons (1.5 tons) of iron ore and 0.303 metric tons (0.33 tons) of coal/coke. During the last 60 years, the scrap industry has prepared, handled, and shipped 1.5436 billion metric tons (1.7 billion net tons) of scrap to U.S. steel mills and foundries. As a result, the use of scrap during this 60-year period has meant conserving nearly 2.3608 billion metric tons (2.6 billion tons) of iron ore and nearly 544.8 million metric tons (600 million tons) of coal/coke — a tremendous savings of virgin resources.

The U.S. Environmental Protection Agency (EPA) has identified six additional benefits when scrap is used instead of virgin materials:[6]

1. 90% savings in virgin materials use
2. 86% reduction in air pollution
3. 76% reduction in water pollution
4. 40% reduction in water use
5. 97% reduction in mining wastes
6. 105% reduction in consumer wastes generated

"Mines above the ground" in America, consisting in late 1975 of over 610.176 million metric tons (672 million tons) of ferrous scrap, (enough to blanket all of Washington, D.C. with nearly 30.48 m [100 ft] of iron and steel discards) are rich enough to supply domestic and foreign-purchased scrap needs (at 1975 levels) for the next 14 years.[5] This figure does not include the huge increment added by every obsolete auto, washing machine, and other metal cast-off which has appeared since the beginning of 1976. The "horn of waste" keeps overflowing at the rate of more than 11.804 million metric tons (13 million tons) per year (Figure 1).[7] Such findings prove that earlier allegations of upcoming scrap shortages did not scientifically calculate the real reservoir level. In contrast to so many of our natural resources, ferrous scrap renewal is virtually unlimited.

By maintaining a status quo national policy, the U.S. at once runs the risk of smothering in its own debris and wasting an energy- and mineral-intensive commodity. As of end-of-year 1975, in fact, 90.8 million metric tons (100 million tons) of ferrous scrap had corroded to the point of uselessness. That process will continue inexorably as long as tax and freight rate structures are skewed in favor of virgin ore, as long as zoning laws inhibit recycling, and until automobile detitling regulations are amended to eliminate that obstacle to maximum recovery of obsolete metal. The policies of today make mining ore domestically, and importing it from the world over, more economically attractive than using the ferrous scrap existing within the U.S. borders.

Iron ore is both finite and energy-voracious. The 90.8 million metric tons (100 million tons) of scrap that are now rusted could have saved the energy-equivalent of 53 billion ℓ (14 billion gal) of gasoline had they been used in place of ore in the steelmaking process. The 65.376 million metric ton (72 million ton) reservoir, if utilized, would reap energy savings equal to over 334 billion ℓ (89 billion gal) of gasoline.

III. SOLID WASTES

By the 1960s it was abundantly clear that recycling had to be applied in a broader context. The nation was throwing away valuable resources due to a lack of recycling incentives and technology.

In the 1967 to 1978 period, 172.52 million metric tons (190 million tons) of the major manufactured materials — paper, metals, glass, textiles, and rubber — were consumed yearly.[8] In the same period, 43.584 million metric tons (48 million tons) of these same materials were recycled through the market annually. Recycled materials were either fabrication wastes or obsolete, discarded products returned to industry for reprocessing. The rate of recycling in the period was 25.2% of consumption. Table 1 presents the recycling rates by commodities.

Solid waste from residential, commercial, and institutional sources amounts to more than 1089.6 kg (2400 lb) per person annually and presents the greatest recycling problem. Fifty percent of this waste consists of paper and wood products, with the balance

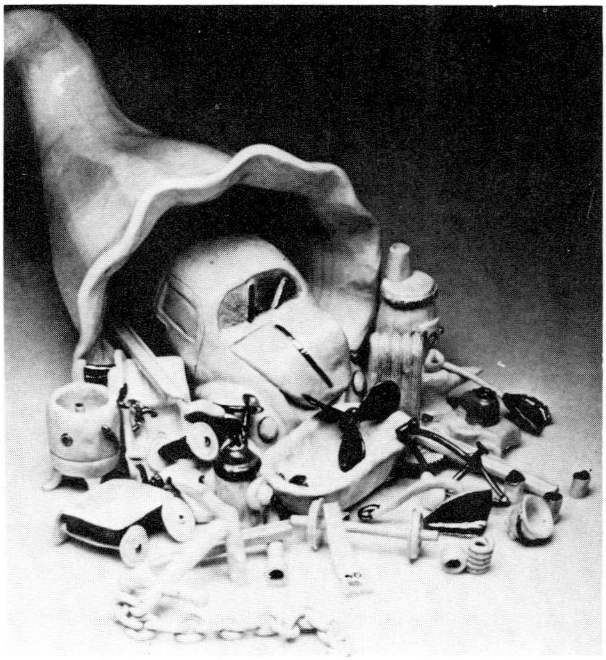

FIGURE 1. Horn of waste.

Table 1[13]
RECYCLING RATES BY COMMODITIES

	Total consumption (million metric tons)	Tons	Total recycled (million metric tons)	Tons	Recycling as percent of consumption
Paper	47.8	53.1	9.1	10.1	19.0
	48.2				
Iron and	95.3	105.9	29.8	33.1	31.2
steel	96.1		30.5		
Aluminum	3.6	4.0	0.6	0.7	18.3
Copper	2.6	2.9	1.3	1.4	49.7
			1.27		
Lead	1.2	1.3	0.5	0.6	49.6
	1.18				
Zinc	1.4	1.6	0.2	0.2	12.6
			0.18		
Glass	11.5	12.8	0.5	0.6	4.2
	11.6				
Textiles	5.1	5.7	0.3	0.3	4.3
	5.17		0.27		
Rubber	3.5	3.9	0.9	1.0	26.2
Total	172.0	191.2	42.7	48.0	25.2
	173.3		43.8		

consisting largely of glass, metal, fibers, plastics, and miscellaneous types of material. These materials are found in such items as bottles, cans, worn-out automobile tires, junk automobiles, newspapers, cartons, and lumber. Table 2 presents a tonnage summary of the solid wastes presented in the U.S. in 1969.

Table 2
SOLID WASTES PRODUCED IN THE U.S. IN 1969[13]

	Million metric tons	Million tons
Residential, commercial, institutional waste (collected, 190 million tons; uncollected, 60 million tons)	225	250
Industrial wastes	99	110
Mineral wastes	1530	1700
Agricultural wastes	2052	2280
Total	3906	4340

Table 3
COMMODITY RECOVERY RATES IN U.S. IN 1970[13]

Commodity	Recovery in percent of U.S. consumption for 1970
Aluminum and alloys from purchased scrap	20% (4% from old scrap)
Antimony	61%
Barite	Only small quantity
Beryllium	0.5%
Bismuth	Small quantity
Cadmium	200,000 lb annually
Chromium	13%
Cobalt	< 1%
Columbium	< 1%
Copper	24%
Gold	29% for 1971
Iron and steel scrap	33 million short tons; 28% of total U.S. consumption of iron
Lead	35%
Manganese	< 1%
Mercury	21%
Molybdenum	Some (figures not available)
Nickel	24%
Platinum group metals	25%
Silver	21%
Tantalum	16%
Tellurium	Small quantity
Tin	26%
Titanium	2987 tons of new scrap recovered (percent figures withheld)
Tungsten	3%
Vanadium	Unknown (probably small)
Zinc	5%
Zirconium	< 1%

Presently, some materials are recovered to a comparatively high degree and others at near zero. Table 3 compares the consumption recovery of solid wastes in the U.S. in 1970.

Each year Americans attempt to throw away 8 million automobiles, 27.24 million metric tons (30 million tons) of paper, 48 billion cans, and 26 billion bottles — enough to make a 181.6 billion kg (400 billion lb) heap of trash each year.[9] The daily per capita production of trash is growing — from 2.5 kg (5.5 lb) per individual in 1975 to an estimated 3.6 kg (8 lb) per individual by 1980. With population increase, the throwaways of the nation are mounting in a geometrical progression.

The decade of the 1970s opened with an environmental eruption that was felt throughout the U.S.[10] A shock wave was sent out which demonstrated a concern for the environment by many who had reached the point of complete frustration.

This crescendo led to "Earth Day", and it brought into sharp focus just how Americans, particularly younger Americans, felt about what was happening to their air, water, and land. A "grass roots" uprising arose to forcefully inform everyone that the people wanted more action than had been forthcoming in the cleaning-up of their environment.

On January 1, 1970, President Nixon signed into law the National Environmental Policy Act, which created the Council on Environmental Quality and required each federal agency to prepare a statement of environmental impact in advance of each major action, recommendation, or report on legislation that might significantly affect the quality of the human environment.

In signing this legislation, President Nixon said, "The 1970s absolutely must be the years when America pays its debt to the past by reclaiming the purity of its air, its waters, and our living environment. It is literally now or never."[14]

This federal commitment, as significant as it was, did not reduce the tempo of "Earth Day" because the initiative for environmental action was coming from the people.

During the summer of 1970, still on the crest of public momentum, the Council on Environmental Quality transmitted its first annual report to the Congress, and stated that "maximum recycling and reuse of materials are necessary to reduce the growing volume of solid wastes that must be disposed of."[15]

Two months later, the Congress passed and President Nixon signed into law the Resource Recovery Act of 1970 — waste disposal of the 1960s became resource recovery in the 1970s. This act also created the National Commission on Materials Policy.

By the end of 1970, through the executive reorganization plan, a new government entity was created — the United States Environmental Protection Agency (EPA), and within that agency, the Office of Solid Waste Management Programs.

In one year, the nation was exposed to the most far-reaching series of actions dealing with environment and solid waste in its history. It was a time of positive anticipation for those who were committed to the benefits and necessity of increased recycling.

Under the Resource Conservation and Recovery Act of 1976, an interagency Resource Conservation Committee, the first cabinet-level group to study the national implications of solid waste policy, was created.[11] Headed by Administrator of the Environmental Protection Agency, it consists of the Secretaries of Commerce, Interior, Labor, and Treasury, the Chairperson of the Council on Environmental Quality, and a representative from the Office of Management and Budget. At 6-month intervals and concluding in April 1978, the committee presented the President and Congress with results of its investigation into the effects of various economic incentives and disincentives and product alterations on resource conservation; ten million dollars had been authorized for the work of the committee.

Areas specifically mandated for study were subsidies, tax structure, restriction of the manufacture or use of certain goods, and the imposition of disposal charges on consumer products, the last of which President Carter had highlighted for accelerated review.

IV. CRYOGENIC RECYCLING OF SOLID WASTES

Cryogenics, a branch of physics that relates to the production and effects of very low temperatures, was born on a Christmas Eve, 1877, when two communications were read before the French Academy of Science.[12] It was here that a scientist by the name of Louis Carltelet first made it known that he succeeded in liquefying oxygen and carbon monoxide by the expansion of gases from high pressures. Simultaneously, Raoul Pictet, an engineer, made his announcement that he had succeeded in liquefying oxygen through mechanical refrigeration. Nitrogen was first liquefied by Wrobleski and Olzewski in 1883. In the words of Edward M. Blanchard, "The Second Century of Cryogenics" began on Christmas Eve in the year of 1977. The cryogenic principle has been toyed with for many years in an effort to accomplish a multitude of tasks.

The use of cryogenics for recycling certain solid waste materials appears to be a relatively new solution to an age-old problem.

The concept for freezing to enhance recycling of solid wastes was initiated by a group of professors in the College of Engineering at the University of Wisconsin in the late 1960s and early 1970s. In an effort to aid the auto salvage dealers in solving the problem of derelict and abandoned automobiles, the problem of the automobile/truck tire surfaced. The tire is not biologically degradable, will not stay buried in landfills, and can no longer legally be burned in the open.

Of course none of the above results in the actual recycling of materials, only in their disposition. The tire of today has been manufactured since 1894 and is accumulating at the rate of 234 million annually, approximately one for each man, women, and child within the U.S. There are literally billions of discarded automobile/truck tires spread throughout the U.S. Many are piled in swamps and caves or just scattered throughout such places as in ditches and the woods. An average discarded nonmetal automobile tire consists of 12 lb of rubber, 6 lb of nylon/rayon, and 1 lb of steel. To multiply these figures by the numbers of discards over the past century results in staggering figures. It is not only a nuisance but has been referred to as black gold, with almost twice its weight in oil used to produce each tire.

It has only been since the early 1970s that interest has developed in cryogenic processing of such difficult-to-reprocess items as the tire.

It has been demonstrated that rubber tires after freezing* −62.2°C (−80°F) can be shattered into a powder and recycled.

From the tire application of cryogenics, technology spread rapidly to other difficult items such as plastic-coated copper wires.

The cryogenic processing of plastic-coated cables involves embrittling the plastic coating by reducing its temperature and impacting the cable. Upon impacting, the brittle plastic will shatter into small particles while the ductile copper wire remains in its original form. Finally, the plastic and wire are separated by conventional techniques.

The cryogenic process not only yields a high grade of scrap but also eliminates the noxious vapors and the need for costly scrubbers. A plastic which can be reused is also produced.

It should be noted that copper is not a cryogenically recyclable material; however, the plastic sheath is. Through this process the whole cable can be recycled.

A problem which hinders recycling is the lack of thought given at the time of manufacturing. For example, bumpers on automobiles are usually made from a steel alloy plated with a coat of copper which in turn is electroplated with chromium. The steel

* Freezing is defined as the process of lowering the temperature of a material until its properties become embrittled.

is functional, the copper is necessary to get the chromium to adhere, and the chromium is solely for customer appeal (cosmetics). Imagine how much simpler the recycling of the automobile bumper would be if it were made from only one material. There are many items being manufactured in a similar manner.

REFERENCES

1. **Hale, S., Jr.,** Salvage markets, SW-29c, U.S. Environmental Protection Agency, Washington, D.C., 1972, 1.
2. **Landau, B.,** Evolution of an industry, *Phoenix Q.,* 7(4), 1, 1976.
3. **Hale, S., Jr.,** Salvage markets, SW-29c, U.S. Environmental Protection Agency, Washington, D.C., 1972, *xvii.*
4. **Porter, S.,** Recycling stymied by inaction, *Wisconsin State Journal,* p. 6, August 27, 1978.
5. **Schapiro, J. S.,** Mines above the ground, *Phoenix Q.,* 9(3), 1977.
6. Reclamation Conservation, Beautification, *Phoenix Q.,* no date.
7. **Schapiro, J. S.,** The horn of plenty keeps overflowing, *Phoenix Q.,* 9(3), 1, 1977.
8. **Hale, S., Jr.,** Salvage markets, SW-29c, U.S. Environmental Protection Agency, Washington, D.C., *xvii.*
9. **Nelson, G.,** Our disposable society: the cost, *Phoenix Q.,* 2(3), 1, 1971.
10. **Cuttler, H.,** The seventies: decade for discussion, *Phoenix Q.,* 7(2), 1, 1975.
11. **Schapiro, J. S.,** Cabinet-level committee surveys conservation options, *Phoenix Q.,* 7(2), 11, 1975.
12. **Vance, R. W.,** *Cryogenic Technology,* John Wiley & Sons, New York, 1963, 293.
13. **Landau, B.,** Energy, material shortages challenge the American way of life, *Phoenix Q.,* 6(1), 4, 1974.
14. **Kirschten, Dick,** President Nixon's remarks when signing the National Environmental Policy Act (NEPA) into law, San Clemente, California, January 1, 1970, *Phoenix Q.,* 11(4), 10, 1980.
15. **Cutler, H.,** The seventies: decade for decision, *Phoenix Q.,* 7(2), 5, 1975.

CHAPTER 2

PROPERTIES OF CRYOGENICALLY RECYCLABLE MATERIALS

N. Braton

TABLE OF CONTENTS

I. INTRODUCTION

The use of cryogenics, a branch of physics relating to the effects of very low temperatures, is having significant impact on the problem of recycling solid wastes. This chapter deals with how a material is determined to be cryogenically recyclable. At the conclusion, applications of cryogenic processing are discussed. (Case studies of cryogenic processing are presented in Chapter 10.)

Basically, all materials to be recycled can be classified into three types: metals, ceramics, and polymers (Figure 1). The fundamental principle of the cryogenic process is to embrittle the material by simple cooling. The embrittled material will then fracture easily, permitting simple physical separation of material from the components.

In order to be classified as cryogenically recyclable, a material must therefore be able to be fractured in a brittle manner. ''Brittle'' meaning fracturing with no apparent plastic deformation.

Some materials will be brittle at room temperatures. Others have to be cooled far below ambient temperatures before they can be fractured in a brittle manner. Still others are ductile at both room and reduced temperatures.

To investigate brittle fracture and the factors affecting brittle fracture, it is convenient to study the yield stress and fracture stress behavior of materials. Figure 2A is a schematic showing the yield and fracture stress behavior of a material which would fracture in a ductile manner, yielding before fracture at both high and low temperatures. This type of material would not be classified as cryogenically recyclable.

Figure 2B is a schematic showing the yield and fracture stress behavior of a material which is ductile at room temperature but brittle at some lower temperature. This type of material is said to go through a ductile-to-brittle transition at T_B, the Brittle Temperature. Below T_B, the material would fracture in a brittle manner, fracturing before yielding. Thus this type of material could be classified as cryogenically recyclable.

Materials which fracture in a brittle manner at room temperature have the room temperature lower than T_B and can also be classified as cryogenically recyclable.

Temperature is not the only factor which affects the yield and fracture behavior of materials. Increasing the rate of loading has the same effect on the yield and fracture stresses as lowering the temperature. Figure 3 demonstrates the effect of increasing the rate of loading on the ductile-to-brittle transition temperature. The increase in T_B due to increasing the rate of loading causes much confusion when trying to quote a particular value of T_B. A value for T_B is only valid for a particular type of test and a particular rate of loading. Unfortunately, very few researchers provide this type of information when publishing values of T_B. This makes comparing different brittle point data difficult. Figure 4 shows how the rate of loading affects T_B for tungsten, zinc, polymethylmetharylate (PMMA) and polyvinyl chloride (PVC).

Many of the cryogenic recycling processes depend on the selective embrittlement of materials. For instance, the cryogenic recycling of plastic-coated copper cables relies on the plastic embrittling and the copper wire remaining ductile. Figure 5 depicts the yield and fracture behaviors of ductile copper wire and a plastic coating which embrittle at some temperature.

The selective embrittlement of materials can also be used to recycle two materials which embrittle if the respective brittle temperatures are not close together. Figure 6 depicts this situation. How great a difference is needed between the brittle temperatures before the selective embrittlement can be effectively used is not known at this time.

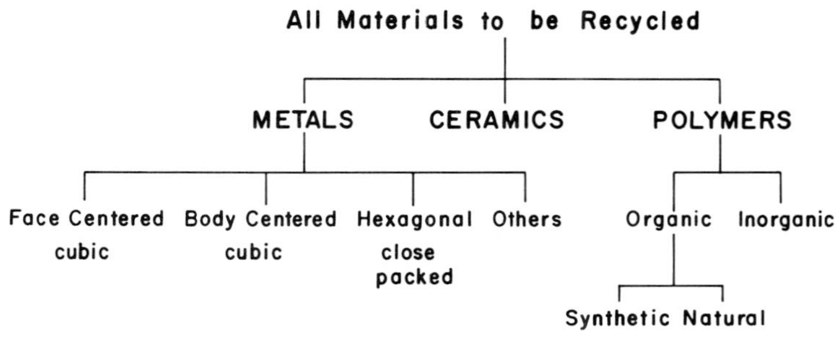

FIGURE 1. Materials classification. (Courtesy of Michalski, P. H., A Classification of Materials Which Can Be Cryogenically Recycled, M. S. thesis, University of Wisconsin—Madison, 1976, 24.)

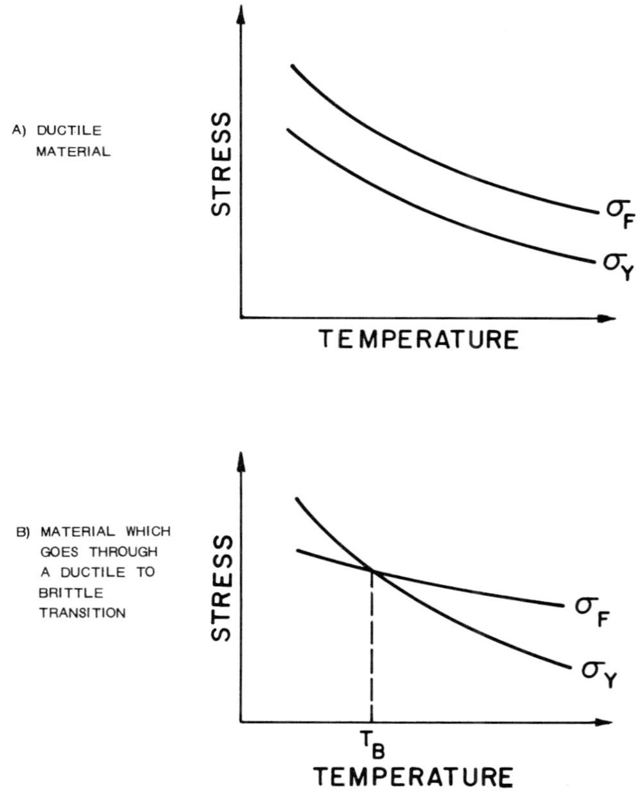

FIGURE 2. Yield stress and fracture stress of (A) a ductile material and (B) a material which goes through a ductile-to-brittle transition. (Courtesy of Michalski, P. H., A Classification of Materials Which Can Be Cryogenically Recycled, M. S. thesis, University of Wisconsin—Madison, 1976, 16.)

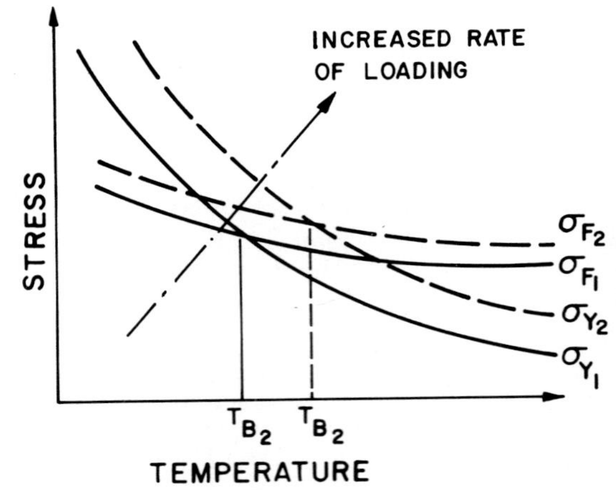

FIGURE 3. Effect of increasing the rate of loading on the ductile-to-brittle transition. (Courtesy of Michalski, P. H., A Classification of Materials Which Can Be Cryogenically Recycled, M. S. thesis, University of Wisconsin—Madison, 1976, 15.)

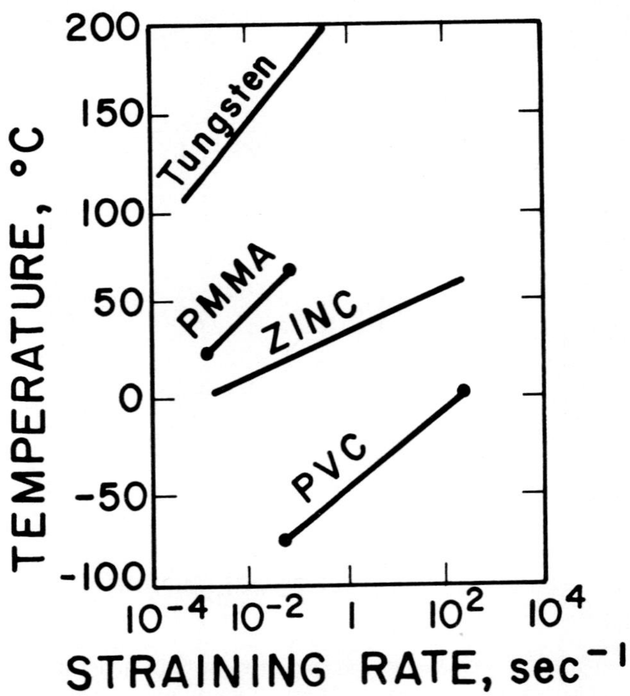

FIGURE 4. Brittle temperatures for tungsten, zinc, PMMA, and PVC for different rates of loading. (Andrews, E. H., *Fracture in Polymers,* Oliver & Boyd, 1968, 50.)

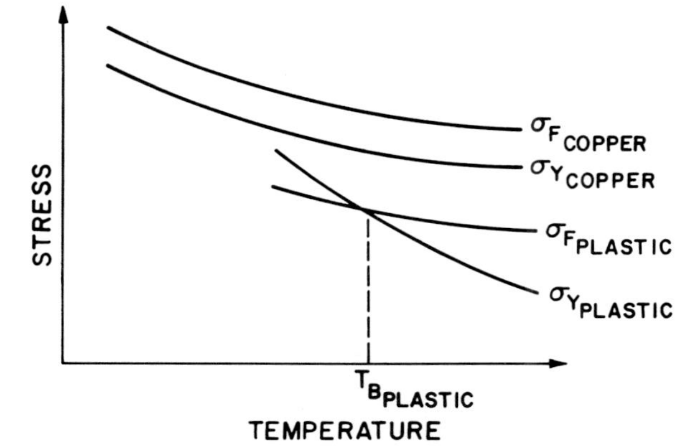

FIGURE 5. Yield and fracture behavior of materials which can be recycled utilizing the selective embrittlement of materials. (Courtesy of Michalski, P. H., A Classification of Materials Which Can Be Cryogenically Recycled, M. S. thesis, University of Wisconsin—Madison, 1976, 16.)

II. SOLID WASTES — MATERIALS CLASSIFICATION

Materials are classified into three groups: metals, polymers, and ceramics. Figure 1 illustrates these groups and their subdivisions. Ceramics, however, have natural properties of brittleness; they are not affected by reduced temperature and therefore will not be discussed in this book.

A. Metals

Of the 103 elements found by man, 86 are metals. Of these, only a few are used in significant quantities as engineering, structural, or electrical materials. They are iron (Fe), copper (Cu), zinc (Zn), aluminum (Al), tin (Sn), platinum (Pt), silver (Ag), gold (Au), lead (Pb), and titanium (Ti). These base materials are often alloyed with other elements to provide a variety of materials, each with their own unique properties.

There are several ways in which metals can be grouped. However, when describing their mechanical properties, it is most convenient to group them according to their crystal structure. Metals have three main crystal structures: face-centered cubic (FCC), body-centered cubic (BCC), and hexagonal close-packed (HCP). Figure 7 shows the atomic structures of each group. Most metals belong in one of these groups and are listed in Table 1. Each crystallographic structure has its own peculiar mechanical properties which allow it to remain ductile or become brittle at low temperatures.

To understand why metals of different crystal structures behave the way they do, two approaches can be taken: microscopic or macroscopic. The microscopic theories developed for slip and cleavage fracture of metals or dislocations in polycrystals will not be discussed here. Rather, the macroscopic approach will be taken, in which the unique mechanical properties for each crystallographic group will be surveyed to establish their ductile or brittle characteristics.

1. FCC Metals

The face-centered cubic (FCC) metals are typically ductile down to −196°C (−320°F), the liquid nitrogen temperature. For this reason, some of the FCC metals

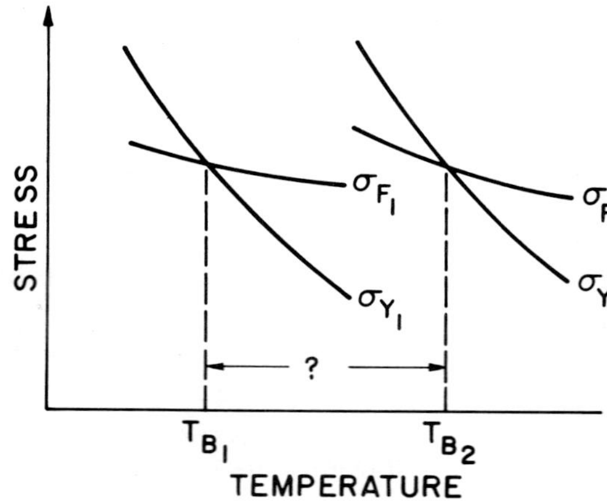

FIGURE 6. Yield and fracture behavior of materials which embrittle at different temperatures. (Courtesy of Michalski, P. H., A Classification of Materials Which Can Be Cryogenically Recycled, M. S. thesis, University of Wisconsin—Madison, 1976, 16.)

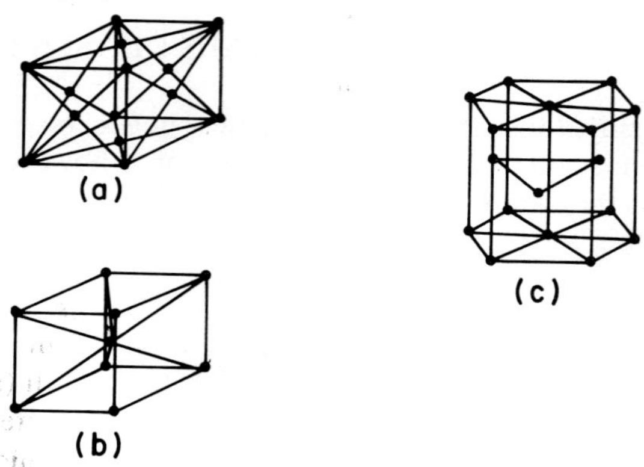

FIGURE 7. Crystallographic structures: (a) face-centered cubic, (b) body-centered cubic, (c) hexagonal close-packed. (Courtesy of Michalski, P. H., A Classification of Materials Which Can Be Cryogenically Recycled, M. S. thesis, University of Wisconsin—Madison, 1976, 27.)

are used in cryogenic applications such as storage tanks for liquid cryogens. To understand why the FCC metals are considered ductile, examination of data obtained from two American Society for Testing and Materials tests is necessary: an impact test and a uniaxial tension test. The temperature-dependent curves of yield strength and tensile strength can be obtained from a series of isothermal tension tests. Similarly, isothermal impact tests can be used to obtain energy absorbed vs. temperature curves.

There are two instructive methods for presenting the tensile test data. Most of the published data on metals have the form of Figure 8 which shows a typical set of curves for an FCC metal. The second is shown in Figure 9.

Several FCC characteristics can be seen in Figure 8. The yield strength (YS) changes little when the temperature is decreased, while the tensile strength (TS) increases significantly. Thus the TS-to-YS ratio (TS/YS) increases as the temperature is decreased.

Table 1
CRYSTALLOGRAPHIC GROUPS

FCC	BCC	HCP
Aluminum (Al)	Chromium (Cr)	Beryllium (Be)
Copper (Cu)	α-Iron (Fe)	Cadmium (Cd)
Gold (Au)	Molybdenum (Mo)	Magnesium (Mg)
γ-Iron (Fe)	Tantalum (Ta)	Titanium (Ti)
Lead (Pb)	Tungsten (W)	Zinc (Zn)
Nickel (Ni)	Vanadium (V)	

Courtesy of Michalski, P. H., A Classification of Materials Which Can Be Cryogenically Recycled, M.S. thesis, University of Wisconsin—Madison, 1976, 59.

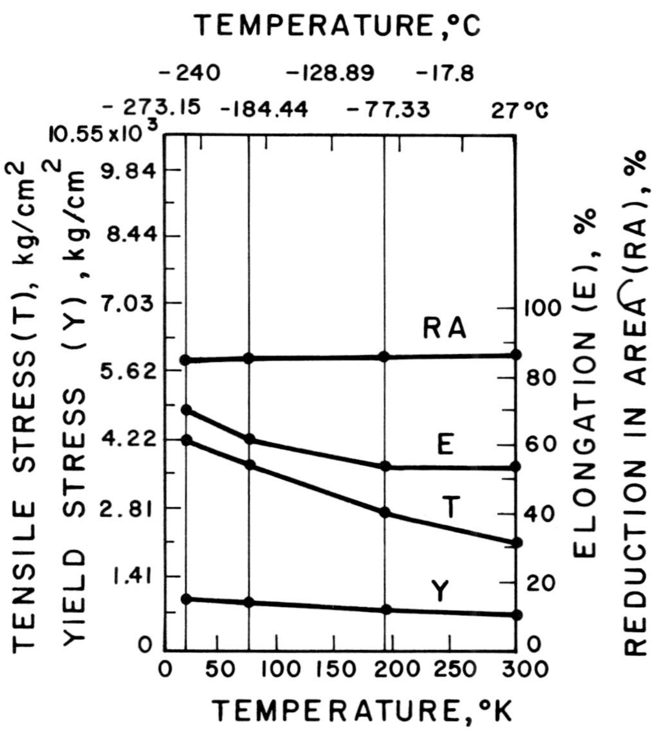

FIGURE 8. Typical FCC tensile test curve. (From Warren, K. A. and Reed, R. P., Tensile and Impact Properties of Selected Materials From 20 to 300 K, Monogr. Ser. No. 63, National Bureau of Standards, U.S. Department of Commerce, 1963, 30.)

This implies that FCC metals are able to accommodate a relatively large amount of plastic deformation before fracture, and also accounts for their extreme reliability at low temperature.

Figure 9 is helpful for understanding the ductile behavior of FCC metals. Notice that the yield strength never touches the ductile fracture strength. This indicates that the FCC metals must plastically deform before they can reach their fracture strength. The crosshatched area is present because Wessel's schematic is indicative of many FCC metals.

An FCC impact test curve is shown in Figure 10. It is typical of a metal which does

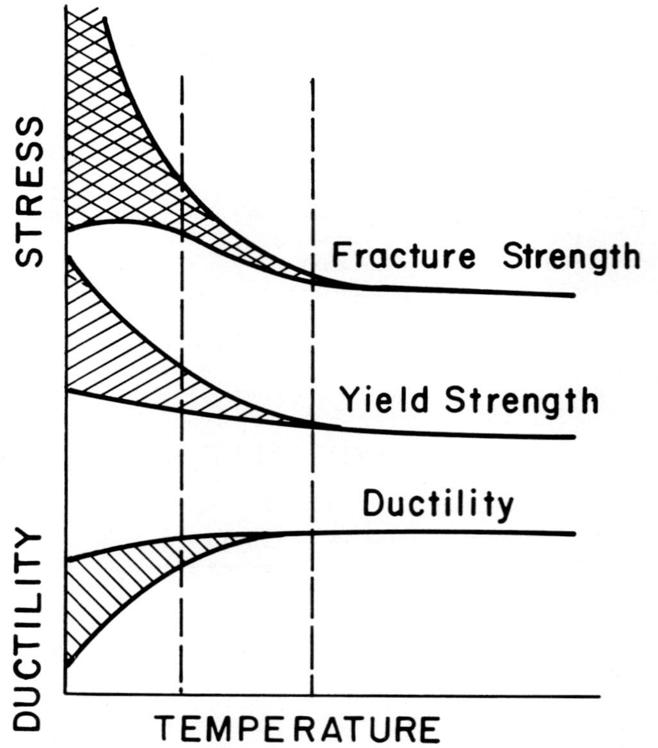

FIGURE 9. Wessel's schematic of FCC metal behavior. (Reprinted with permission from ASTM Special Technical Publication 387, p. 30, copyright American Society of Testing and Materials, Philadelphia, Pa.)

not go through a ductile-to-brittle transition. Notice that the curve is relatively temperature-independent and has a relatively high value compared to a body-centered cubic (BCC) impact curve (Figure 11).

In summary, the FCC metals remain ductile at low temperatures.

2. BCC Metals

Body-centered cubic metals (BCC), unlike the FCC metals, tend to become brittle at low temperatures. To understand why there is this difference, the uniaxial tension tests and impact test will again be examined.

Typical BCC data are shown in Figure 12. There is one noticeable difference between this figure and Figure 8. The yield strength is rather temperature-dependent and rises faster than the tensile strength. This produces a decreasing TS-to-YS ratio when the temperature is decreased. This behavior is typical for a metal which will go through a ductile-to-brittle transition.

Wessel's schematic for BCC metals is shown in Figure 13 and is useful in describing the ductile-to-brittle transition in BCC metals. At high temperatures, the metals behave in a ductile manner, plastically yielding before fracture. However, upon decreasing the temperature, less and less plastic deformation is experienced before fracture occurs. Finally, when the temperature falls to T_1, no plastic deformation occurs prior to fracture.

Another important feature which should be noticed is the ductile-to-brittle transition

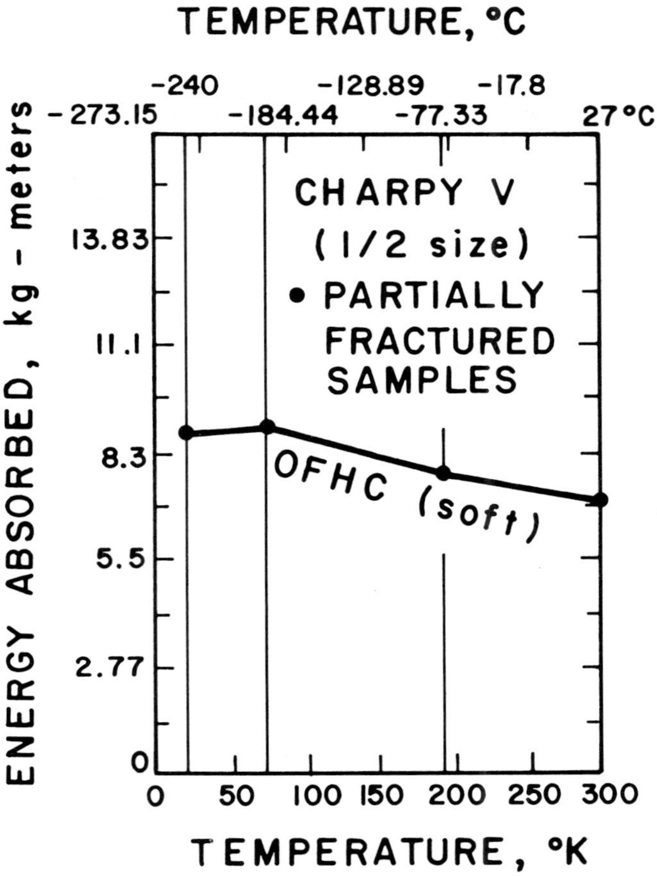

FIGURE 10. Typical FCC impact test curve. (From Warren, K. A. and Reed, R. P., Tensile and Impact Properties of Selected Materials From 20 to 300 K, Monogr. Ser. No. 63, National Bureau of Standards, U.S. Department of Commerce, 1963, 32.)

range. There is a range of temperatures in which a material could fracture in a brittle manner. This range is due to the rather pronounced differences in fracture strength that can be observed depending on the metal, its metallurgical condition, and the testing temperature. The nondimensional temperature abscissa of Figure 13 is s convenient means of drawing a schematic for many materials.

Figure 11 is representative of BCC impact data. The sharply falling impact values are typical of brittle metals. In summary, commercially pure BCC metals will become embrittled at low temperatures.

3. HCP Metals

The hexagonal close-packed (HCP) metals have properties which lie between the ductile FCC metals and the embrittling BCC metals. Some of the HCP metals will remain ductile at low temperatures and some will not. For instance, cadmium, which remains ductile down to the lowest temperatures, is very soft and liable to creep, while zinc is highly brittle.

The mechanical properties of other HCP metals such as titanium and beryllium are greatly affected by interstitial impurities such as oxygen, hydrogen, and nitrogen. Increasing the interstitial impurity content will increase the yield strength and tend to

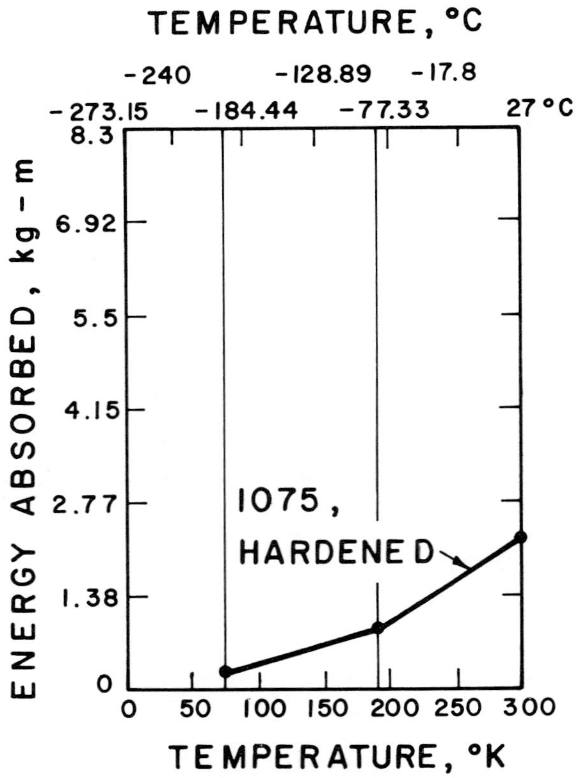

FIGURE 11. Typical BCC impact test curve. (From Warren, K. A. and Reed, R. P., Tensile and Impact Properties of Selected Materials From 20 to 300 K, Monogr. Ser. No. 63, National Bureau of Standards, U.S. Department of Commerce, 1963, 36.)

promote brittle fracture. Pure titanium, however, is an attractive material for cryogenic purposes because as the temperature is reduced, the TS-to-YS ratio remains constant, or even increases.

Steel, which is mainly iron (Fe), can exist in either an FCC structure or a BCC structure. The differences in structure can be obtained by alloying and heat-treating. The most widely used FCC steels are the austenitic stainless steels of the 300 series. The BCC steels are called ferritic steels (α-Fe).

So much has been published regarding the mechanical properties of ferrous materials that it would be impractical to condense it at this writing. However, the Society of Automotive Engineers has published a special publication, *Low Temperature Properties of Ferrous Materials,* which is excellent.

There are two exceptions to the preceding general discussion of the mechanical properties of metals. The mechanical properties of two copper alloys differ from those of typical FCC metals. First, there is a copper-antimony alloy which is brittle at low temperatures. Second, there is a manganese-bronze alloy which has impact strengths that decrease as the temperature is raised. This decrease in impact strength is due to grain boundary precipitation at elevated temperatures.

Table 2 is a classification of the metals tested.

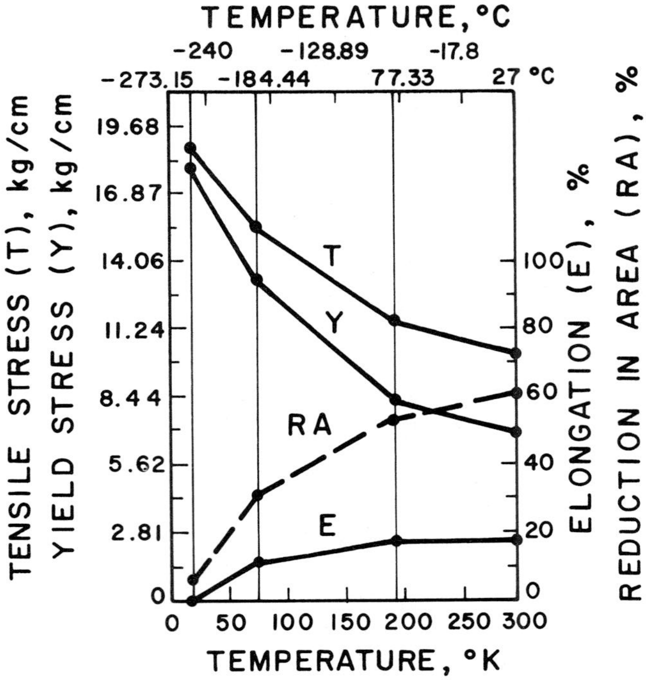

FIGURE 12. Typical BCC tensile test curves. (From Warren, K. A. and Reed, R. P., Tensile and Impact Properties of Selected Materials From 20 to 300 K, Monogr. Ser. No. 63, National Bureau of Standards, U.S. Department of Commerce, 1963.)

B. Polymers

Polymers are large molecules made from many (poly-) smaller repeat units (mers) which are held together by strong primary bonds called covalent bonds. An example of a repeat unit is

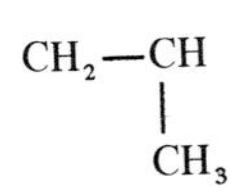

which, together with many other repeat units, forms polypropylene.

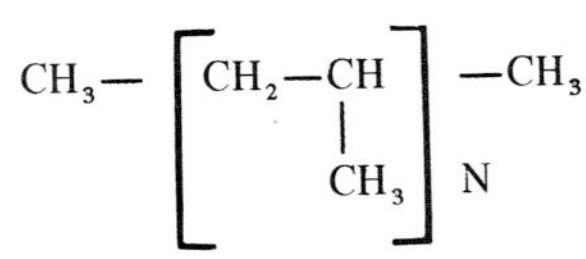

The structural formula for polypropylene demonstrates one of the differences between metals and polymers. Notice that polypropylene can have many different molecular weights depending upon the value of N, whereas, a metal has but one molecular weight. Polymers typically have molecular weights between 10^4 and 10^6, while the highest molecular weight metal is Lawrencium (Lw) with a molecular weight of 103.

Polymers can be divided into two main categories: thermosets and thermoplastics. The difference between the two lies in the type of bonds which hold adjacent molecules together. In thermosets, the adjacent molecules are held together like repeat units by covalent bonds and form what is called a framework, or network, of molecules. Thermosetting polymers are said to be crosslinked by covalent bonds. On the other hand, thermoplastics molecules are not crosslinked but are held together by weaker secondary bonds called van der Waals' forces.

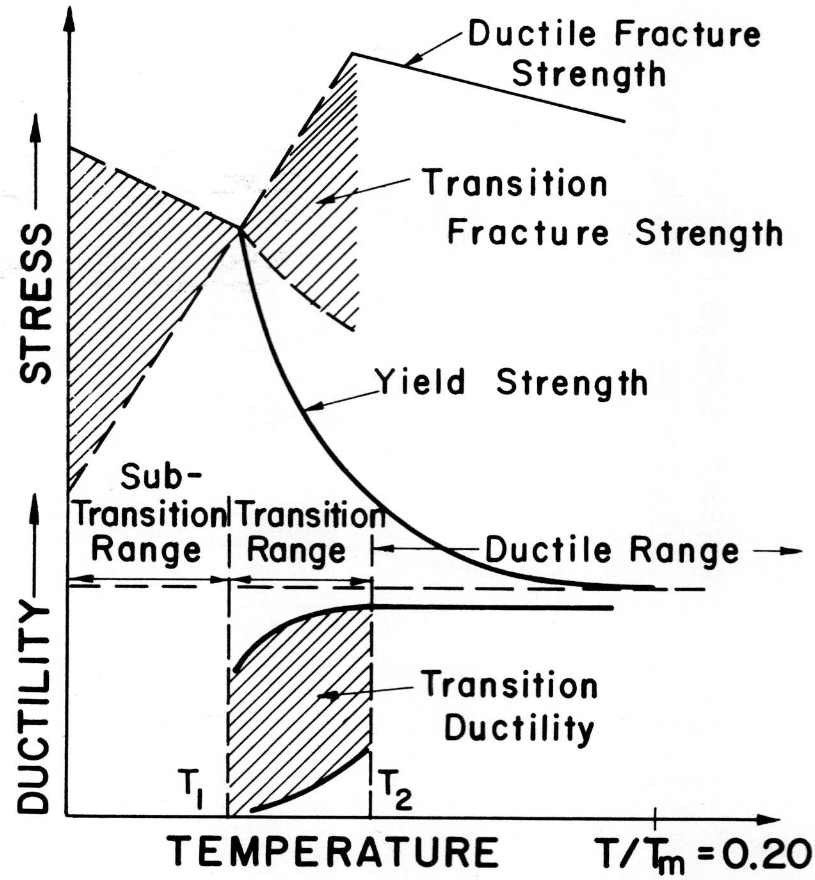

FIGURE 13. Wessel's schematic of BCC metal behavior. (Reprinted, with permission, from ASTM Special Technical Publication 387, p. 35, copyright American Society for Testing and Materials, Philadelphia, Pa.)

Table 2
CLASSIFICATION OF METALS

Materials	Source	Brittle at 20°C	Brittle at −196°C	Ductile at −196°C
Aluminum				
Cast 356	SIGNICAST[a]			X
Cast QQ-A-601	Falk[a]			X h
Rounds 2011-T3	RRSN[a]			X
Rounds 2017-T4	RRSN			X
Rounds 2024-T4	RRSN			X
Rounds 6061-T6	RRSN			X
Copper				
OFHC electrical cane				X
Cast bronze 8-10-10	SIGNICAST			X
Cast silicon bronze	SIGNICAST			X
Bronze SAE 65	Falk			X h
Bronze SAE 660	Falk			X h
Bronze SAE 660	Falk			X h
Brass ASTM B21 (AMPCO brass)	Falk			X h

Table 2 (continued)
CLASSIFICATION OF METALS

Materials	Source	Brittle at 20°C	Brittle at −196°C	Ductile at −196°C
Iron				
Steel				
CR 1018	RRSN		X	
SAE 1020	Falk		X	
HR 1020	RRSN		X	
SAE 1045	Falk			X h
CR 1045	RRSN		X	
HR 1045	RRSN			X
Mild steel water pipe	—		X	
SAE 6155 (chrome vanadium)	Falk			X h
Stainless steel				
17-4 PA	RRSN			X
203 EZ	RRSN			X
303	RRSN			X
304	RRSN			X
316	RRSN			X
410	RRSN		X	
Stainless steel 416	RRSN			X
Cast iron				
Ductile	G&L[a]		X h	
Ductile (fully annealed)	Chromfed.[a] Mal.[a]	X h		
Ductile (80-55-06)	Chromfed. Mal.		X h	
Grey	Mil. Mal.[a]	X h		
Grey	Falk		X h	
Malleable	Mil. Mal.		X h	
Cast steel				
As cast #1	Falk			X h
As cast #1 (quenched and tempered)	Falk			X h
As cast #2	Falk			X h
Cast ASTM-A148	Falk			X h
Pelton P3	Pelton[a]		X h	
Pelton P4	Pelton		X h	
Pelton P10	Pelton			X h
Cast 4140	SIGNICAST		X	
Cast 8620	SIGNICAST		X	
Cast stainless				
SS 316	SIGNICAST			X
SS 25-5PH	SIGNICAST			X
Lead base babbit	Falk	X h		
Tin base babbit	Falk		X h	

[a] Industry supplying materials for the test.

Courtesy of Michalski, P. H., A Classification of Materials Which Can Be Cryogenically Recycled, M. S. thesis, University of Wisconsin—Madison, 1976, 42.

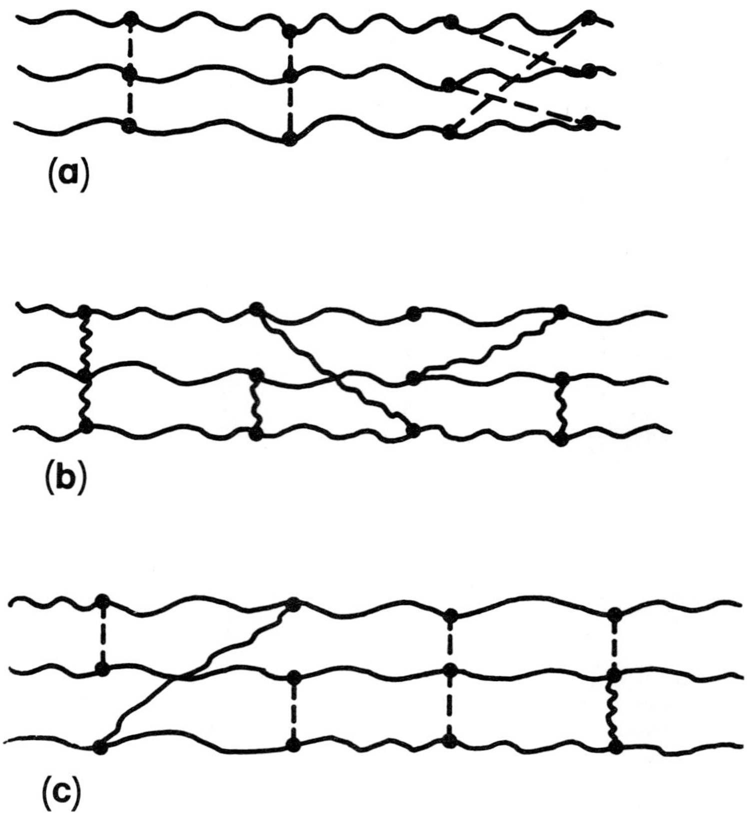

FIGURE 14. A schematic showing the differences in bonding between thermoplastics: (a) three molecules held together by van der Waals' forces (dotted lines) forming a thermoplastic, (b) a network of covalent bonds (solid lines) forming a thermoset, (c) a lightly cross-linked polymer with both van der Waals' forces and covalent bonds forming an elastomer. (Courtesy of Michalski, P. H., A Classification of Materials Which Can Be Cryogenically Recycled, M. S. thesis, University of Wisconsin—Madison, 1976.)

The amount of crosslinking has led to some confusion. Polymers which have many crosslinks are commonly called thermosets, but polymers with only a few crosslinks have come to have two names. Several authorities refer to lightly crosslinked polymers as elastomers, while others refer to them as thermosets. For the purpose of this discussion, lightly crosslinked polymers will be referred to as elastomers. Figure 14 demonstrates the difference between thermoplastics, thermosets, and elastomers.

The dots in Figure 14 are repeat units which are held together by covalent bonds (solid lines). In thermoplastics polymers (Figure 14a) the adjacent molecules are held together by van der Waals' forces (dotted lines). Figure 14b is a schematic of a thermosetting polymer which is highly crosslinked with covalent bonds (solid lines). Figure 14c is a schematic of an elastomer with both covalent bonds and van der Waals' forces holding the polymer together.

The different bonds affect how polymers react at high and low temperatures. Once formed, thermosets and elastomers which are crosslinked polymers will not remelt due to the strong covalent bonds present between molecules. Thermoplastics, on the other hand, can be remelted because only the weak van der Waals' forces hold adjacent molecules together.

Because these polymers differ in behavior at high temperatures, different techniques must be used to recycle them. Once in powder or granular form, thermoplastics can be remelted and reformed into new products. Unfortunately, thermosets and elastomers cannot be remelted, so other means of recycling these materials must be found. For example, ground-up rubber found in automobile tires, an elastomer, is mixed with a binder and used to resurface roads, runways, and flat roofs. Other tires' rubber is being devulcanized (having their crosslinks chemically removed) and remolded. Using ground-up thermosets as fillers for other plastic products is a possibility.

Unlike metals which are entirely crystalline, polymers can take on either an amorphous structure or a crystalline structure when in the solid phase.

Due to their rigid network type of structure, the molecules of thermosetting polymers cannot form a crystalline structure. The movement needed to reorganize the molecules into a crystalline structure is prevented by the covalent bonds which are formed between the molecules when curing.

Thermoplastics polymers, on the other hand, can exist in either an amorphous or a highly crystalline structure with the amount of crystallinity varying between zero and close to 100%. When cooling from the liquid phase, thermoplastics molecules can reorganize themselves into a crystalline structure because only the weak van der Waals' forces are present to restrict the movement of molecules.

Crystallizable elastomers cannot crystallize as freely as thermoplastics due to the small number of crosslinks present and also due to the fact that nearly all elastomers have melting points far below room temperature (i.e., natural rubber has a melting point of −146°C [−230°F]).

It should be noted that no polymer can crystallize 100% and that any portion of the polymer that doesn't crystallize will have an amorphous structure.

In order to crystallize, polymers should have a symmetrical structure. Unsymmetrical or branched molecules will hinder the crystallization process. Examples of symmetrical, unsymmetrical, and branched molecular structures are shown in Figure 15.

Linear polyethylene, which is symmetrical, will crystallize and form high-density polyethylene. The branched polyethylene, however, will remain amorphous or crystallize only a small amount and form low-density polyethylene. Polystyrene, which is unsymmetrical, will form an amorphous solid.

Crystallization also depends upon the rate a polymer is cooled and the manufacturing process used to form the product. Figure 16 shows how cooling and processing affects the structure of thermoplastics.

The mechanical properties of thermoplastics and elastomers are highly dependent on temperature and rate of loading, whereas the mechanical properties of thermosets usually have neither of these dependencies. Even though the mechanical properties of these polymers differ, they can be portrayed on one set of stress-strain curves as shown in Figure 17.

Thermosets are typically rigid polymers which fracture in a brittle manner with stress-strain curves similar to curve (Figure 17a). At different temperatures, thermosets still behave in a brittle manner.

At low temperatures, thermoplastics and elastomers behave like thermosets and fracture in a brittle manner with stress-strain curves similar to curve (Figure 17a). However, at higher temperatures, thermoplastics and elastomers start to yield and cold draw, as shown in curve (Figure 17b). At still higher temperatures, thermoplastics will melt and behave like a fluid with deformation stress proportional to the applied strain rate, as shown in curve (Figure 17d). Elastomers, with their crosslinks preventing fluid flow, will have a rubbery behavior of curve (Figure 17c).

Temperature is not the only factor which influences the mechanical properties of

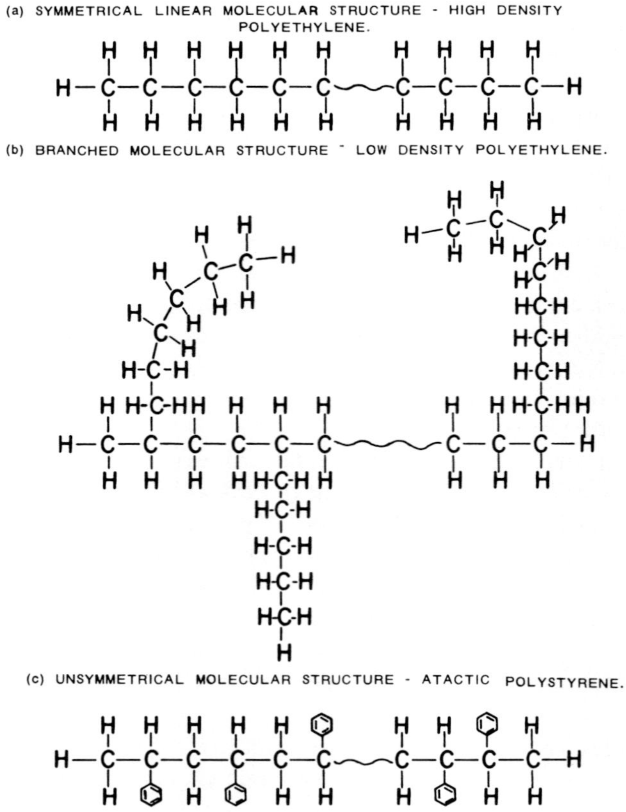

FIGURE 15. Structural formulae for symmetrical and unsymmetrical polymers. (Courtesy of G. Halldin, unpublished, 1980.)

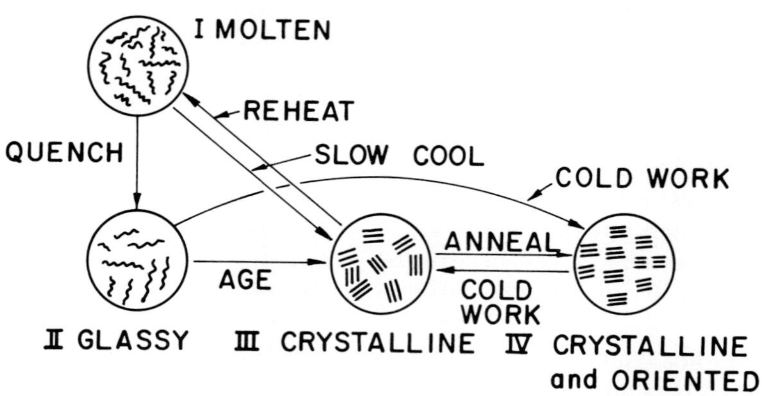

FIGURE 16. Thermal and deformation treatment of polymers. (Van Vlack, L. H., *Elements of Materials Science,* 21E, 1964, Addison-Wesley, Reading, Mass., p. 190. Reprinted with permission.)

thermoplastics and elastomers. A faster rate of loading will affect the mechanical properties of thermoplastics and elastomers in the same fashion as does lowering the temperature (Table 3).

Manufacturing processes also affect the mechanical properties or polymers. For example, nylon-6 molded gears will fracture when impacted in a hammermill at −196°C

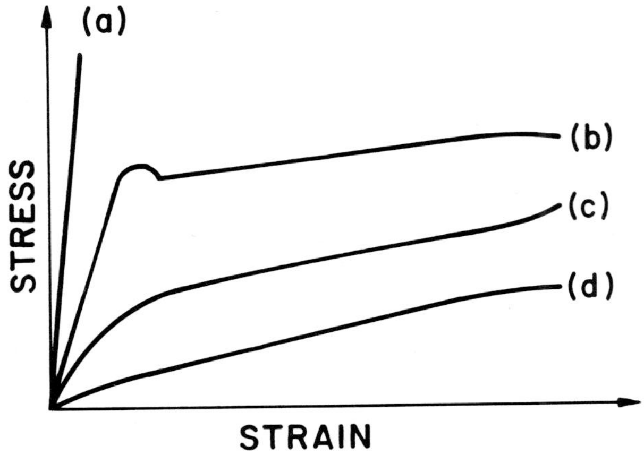

FIGURE 17. Mechanical properties of polymers. (Courtesy of Michalski, P. H., A Classification of Materials Which Can Be Cryogenically Recycled, M. S. thesis, University of Wisconsin—Madison, 1976, 55.)

Table 3
PARAMETERS AFFECTING CHANCE OF BRITTLE FAILURE

Increased factor	Chance of brittle failure
↑ Temperatures	↓
↑ Rate of loading	↑
Molecular property	
↑ M_N	↓
↑ Cross-linking	↑
↑ Crystallinity	↑[a]
↑ Orientation	↑
↑ Length of linear-side chains	↓
↑ Rings or side groups	↑
Additives	
↑ Plasticizers	↓
↑ Fibers	↑↓[b]
↑ Rubbery polymers	↓

[a] Not including films or fibers.
[b] Glass fibers increase yield stress so increase the chance of brittle fracture but when added to brittle resins, they may increase brittle strength and so reduce chance of brittle failure.

Courtesy of Michalski, P.H., A Classification of Materials Which Can Be Cryogenically Recycled, M. S. thesis, University of Wisconsin—Madison, 1976, 59.

(−320°F), whereas 0.05 mm (0.002 in.) thick nylon-6 film will not. This ductile behavior is typical for fibers and films and is caused by the high degree of orientation in these polymers.

Like some metals, some polymers will go through a ductile-to-brittle transition. An explanation of the ductile-to-brittle transition in polymers can be obtained from an understanding of their yield and fracture behaviors.

In a macroscopic sense, the yield and fracture stresses of polymers act independently of one another. They act independently in the sense that they have different magni-

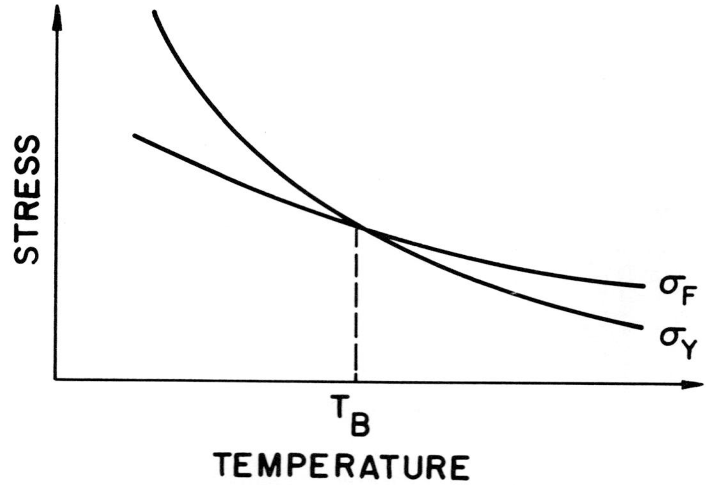

FIGURE 18. Ductile-to-brittle transition in polymers. (Courtesy of Michalski, P. H., A Classification of Materials Which Can Be Cryogenically Recycled, M. S. thesis, University of Wisconsin—Madison, 1976, 57.)

tudes and different rates of change with temperature and rate of loading. Figure 18 demonstrates how the independent nature of these two stresses influences the ductile-to-brittle transition in polymers. Since the yield stress and the fracture stress have different rates of change with temperature, the two stress curves will intersect at some temperature, the ductile-to-brittle transition (T_B). Above T_B the polymer will behave in a ductile manner, yielding before fracture. Below T_B the polymer will behave in a brittle manner, fracturing before yielding.

As mentioned previously, increasing the rate of loading has an effect similar to decreasing the temperature. Figure 19 demonstrates the effect of the rate of loading on the ductile-to-brittle transition of polymers. If the rate of loading is increased, the yield stress will shift from $^{\sigma}y_1$ to $^{\sigma}y_2$ and the fracture stress from $^{\sigma}f_1$ to $^{\sigma}f_2$ with the overall effect of raising T_B and increasing the chance of brittle fracture.

Any factor which raises the yield stress relative to the fracture stress will increase the chance of brittle failure. Any factor which lowers the yield stress relative to the fracture stress will decrease the chance of brittle failure. For instance, if the number, average molecular weight (M_N), is decreased, the fracture stress will decrease, while the yield stress remains the same. The overall effect of this change will be to raise T_B and increase the chance for brittle failure. Figure 20 demonstrates the effect of lowering M_N.

For many common polymers, polyisoprene, polystyrene, and polyisobutylene, the brittleness temperature, T_B, lies close to the glass-transition temperature, T_G. However, this is not always true. For instance, T_G for rigid pure PVC is 74°C (165°F), while T_B is −20°C (−4°F). Table 4 lists T_G and T_B for several polymers. These differences can be attributed to the definitions of T_B and T_G. T_B is defined by the technique employed to fracture a specimen (i.e., notched bar impact test, unnotched bar impact test), whereas T_G is defined by a nondestructive type of test which measures the molecular motion of the specimen.

The notion that if T_B doesn't lie close to T_G, then it lies close to a secondary lower temperature transition, is also a fallacy. Polycarbonate, for example, has its lowest transition at −73°C (−99°F) (at 150 C/S) but is ductile down to −196°C (−320°F) in fracture.

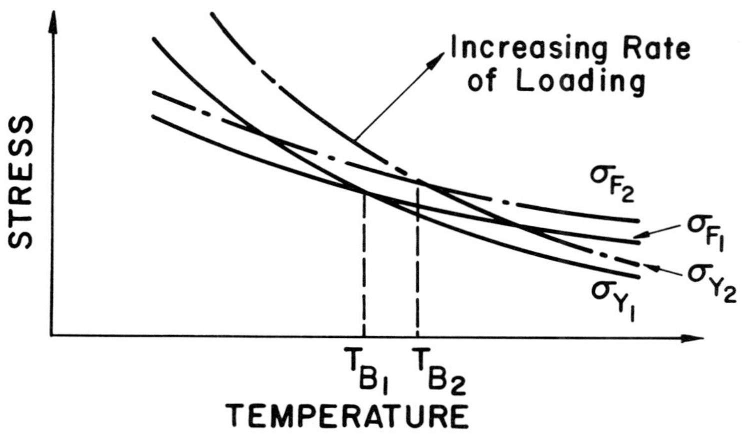

FIGURE 19. Effects of increased rate of loading on ductile-to-brittle transition in polymers. (Courtesy of Michalski, P. H., A Classification of Materials Which Can Be Cryogenically Recycled, M. S. thesis, University of Wisconsin —Madison, 1976, 57.)

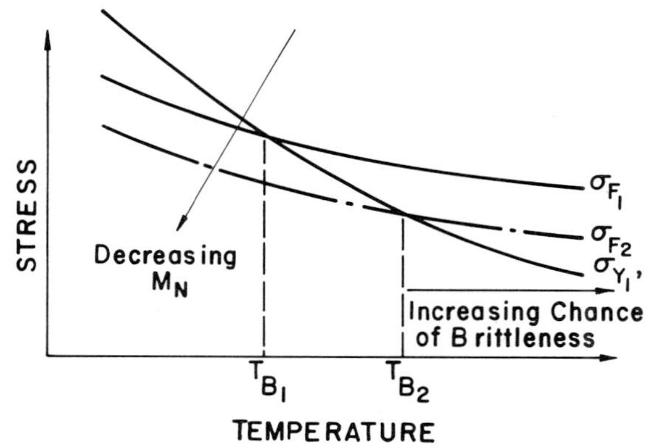

FIGURE 20. Effect of the molecular weight on T_B. (From Andrews, E. H., *Fracture in Polymers,* Oliver & Boyd, 1968, 60. With permission.)

Table 4
GLASS TRANSITION AND BRITTLE TEMPERATURES FOR POLYMERS

Polymer	T_G°C	T_G°F	T_B°C	T_B°F
PMMA	105	221	45	113
Polycarbonate of bisphenal A	150	302	−200	−328
Rigid PVC	74	165	−20	−4
Natural rubber (polyisoprene)	−70	−94	−65	−85
Polystyrene	100	212	90	194
Polyisobutylene	−70	−94	−60	−76

Courtesy of Michalski, P. H., A Classification of Materials Which Can Be Cryogenically Recycled, M.S. thesis, University of Wisconsin—Madison, 1976, 61.

The relationship between T_G and T_B has been stated well by Andrews[19] in that "T_G and T_B are not simply related, but if the yield stress σ_y is increasing rapidly in the region of T_G, as it often does for amorphous polymers, it is likely that σ_y will first exceed σ_f (brittle) in that region, and T_B will then be close to T_G".

Table 5 lists polymers tested at the University of Wisconsin—Madison for their properties at three different temperatures.

III. APPLICATIONS OF CRYOGENIC PROCESSING

Using the general background principles of cryogenic processing, let us now turn toward some specific promising application which might be of industrial importance.

A. The Tire-Disposal/Recycling Problem

One of the largest volume-produced commodities made from synthetic polymeric materials is tires. Presently, about 1.8 billion kg (four billion lb) of tires are being produced annually. Present disposal methods such as landfills, composting, ocean dumping, and burning are inadequate due to environmental considerations. Older methods of rubber recycling, using shredding and separation plus chemical treatment to obtain reclaimed rubber, have only accounted for approximately 10% of the discarded tires. The high cost of shipping whole discarded tires has been the limiting factor in reclaiming rubber.

Cryogenic processing of rubber tires is a simple operation, and the resulting product is in the form of particles which predominantly fall in the 20 to 10 mesh size. The fibrous and wire reinforcements are readily separated.

Utilization of the rubber particles is now the key question. The estimated cost of the rubber particles produced by this method is about 5¢/lb. There are a number of processing routes, as indicated in Figure 21. Table 6 provides a list of suggested specific uses of the recycled rubber particles.

Rubber has been mixed with asphalt binders for road surfaces and been found to have certain favorable characteristics. A 5% mixture of rubber in asphalt has resulted in raising the softening point of asphalt; hence there is less flow at high temperature. It also lowered the brittle point resulting in less cracking at cold temperatures and imparting elastic properties to the asphalt. Further, a rubberized asphalt binder has also been found to better hold the aggregate.

Although adding rubber to asphalt was first tried and patented 130 years ago, accurate records have not been kept.[23] In the 1920s and 1930s, some roads of this material were laid in Europe, with one particular stretch in New Cross, England, lasting until the 1960s. Only recently has interest been rekindled. It is reported that in excess of 16,102 km (10,000 mi) of rubberized roads have been laid. Battelle Institute has published a report on work that they have done for EPA with rubberized asphalt.[19]

The city of Phoenix, Ariz. laid in excess of 174,000 m^2 (208,000 yd^2) of rubberized asphalt on its major streets and airport aprons in the summer of 1971, with far less disruption of traffic and far less cost than with conventional materials. The city expects 10 years of maintenance-free service, all at a substantial savings to the taxpayer.[20]

Recycled rubber particles are utilized by Minnesota Mining and Manufacturing in making a base for athletic fields, tennis courts, and race tracks. General Motors Research Center has found that by mixing 10 to 12% rubber with proportionate amounts of coal, a desirable fuel is developed with 13,500 Btu/lb of mixture.[21] The undesirable hydrocarbons associated with the burning of tires reportedly disappears. Northern States Power Company in Minnesota ran similar tests in 1973 and 1974. Personnel at the University of Wisconsin—Madison have been interested in the direct molding of

Table 5
CLASSIFICATION OF POLYMERS

Materials	Source	Brittle at 20°C	Brittle at −196°C	Ductile at −196°C
Acrylonitrile butadien styrene (ABS)	FLMB[a]		X	
Acrylonitrile butadien styrene (ABS)	WSTRN PLSTCS[a]		X	
Alkyd molded parts	PL[a]	X		
Cellulose acetate butyrate	WSTRN PLSTCS		X	
Fiberglass rod	RRSON[a]	X		
Melamine/phenolic	PL	X		
Melamine/phenolic	Red Rock[a]	X		
Nylon	FLMB		X	
66 Nylon	Bemis[a]		X	
66 Nylon	WSTRN PLSTCS		X	
Nylon filled with molybdenum disulfate	WSTRN PLSTCS		X	
66 nylon plus modifier	G.E.[a]	X		
Phenolic	G.E.	X		
Phenolic	P.L.	X		
Phenolic	P.L.	X		
Phenolic	Red Rock	X		
Phenolic	Red Rock	X		
Phenolic	LAPCOR[a]	X		
Polyacetal	G.E.	X		
Polyacetal	WSTRN PLSTCS	X		
Polyacetal celcon	FLMB	X		
Polybutylene terephalate	LWS[a]	X		
Polycarbonate	G.E.		X	
Polycarbonate	Dstt[a]		X	
Polycarbonate	FLMB		X	
Polycarbonate	WSTRN PLSTCS		X	
Polycarbonate	LWS		X	
Polyester plus glass Fibers	G.E.	X		
Polyethylene — low density	FLMB		X	
Polyethylene — low density	WSTRN PLSTCS		X	
Polyethylene — low density	LWS		X	
Polyethylene — low density	LWS		X	
Polyethylene — high density	WSTRN PLSTCS		X	
Polyethylene — high density	FLMB		X	
Polyisopene (natural rubber)	Burton[a]		X	
Polymethylmethacrylate (PMMA)	G.E.	X		
Polymethylmethacrylate (PMMA)	G.E.	X		
Polypropylene	Dstt	X		
Polypropylene	FLMB		X	
Polypropylene	LWS	X		
Polypropylene	WSTRN PLSTCS	X		
Polypropylene — impact grade	WSTRN PLSTCS		X	
Polypropylene — co-polymer	LWS		X	
Polypropylene — oxide	LWS		X	
Polystyrene	FLIXIB	X		
Polystyrene	WSTRN PLSTCS	X		
Polystyrene	LWS	X		
Polystyrene — medium impact grade	WSTRN PLSTCS	X		

Table 5 (continued)
CLASSIFICATION OF POLYMERS

Materials	Source	Brittle at 20°C	Brittle at −196°C	Ductile at −196°C
Polystyrene — high impact grade	Bemis		X	
Polysulfone	G.E.		X	
Polyurethane	G.E.		X	
Polyvinylchloride	RRSON	X		
Silicon	Burton		X	
Vinyl	FLMB		X	

[a] Companies supplying materials.

Courtesy of Michalski, P. H., A Classification of Materials Which Can Be Cryogenically Recycled, M. S. thesis, University of Wisconsin—Madison, 1976, 63.

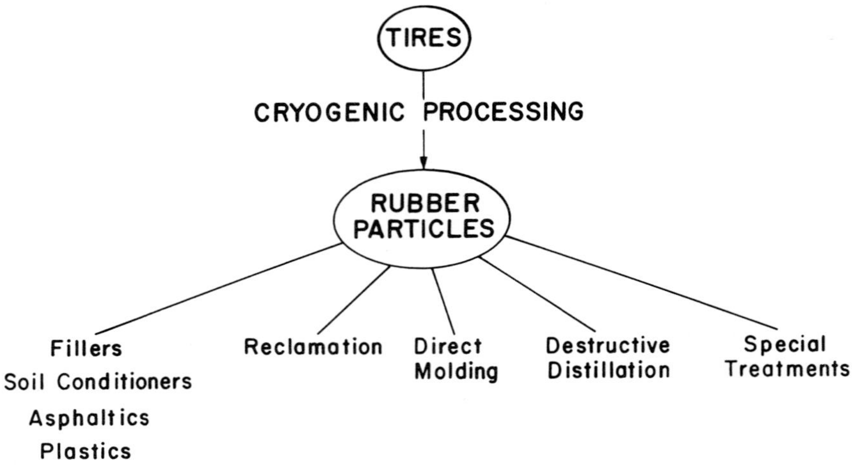

FIGURE 21. Flow chart for cryogenically processing rubber from tires. (From Kisielewski, Robert, Cryogenic Recycling International, La Crosse, Wis., 1977.)

these rubber particles mixed with certain additives such as cellulose fines and synthetic powdered resins, such as polyethylene and polypropylene. A dramatic increase in strength and elongation can be achieved by incorporation of only 10% by weight of polyolefins.

The destructive distillation of these rubber particles might be even more exciting, since now they can be shipped at much less cost to distillation sites. Work has been done in evaluation of the types of yields in the distillation process. The amounts of solid residue, oil, or gas produced depends on the temperature of the distillation process.

Work done by the Firestone Tire and Rubber Company of Akron, Ohio, in conjunction with the U.S. Bureau of Mines, determined that the destructive distillation of scrap tires is technically feasible.[27] The economic feasibility has yet to be determined; cost of shipping the tires to distillation plants is a critical point. In this study, over 568 ℓ (150 gal) of oil, 42.48 m^3 (1500 ft^3) of gas (comparable in heating value to natural gas),and 380 kg (850 lb) of residue per ton of tires were obtained via destructive distillation.

Table 6
SUGGESTED USES FOR RECYCLED RUBBER

Asphalt mixes	
	Road, runways, streets, flat roofs, playgrounds
Fuel	
	Stoker mixes of 10% rubber, 90% coal
Filters	
	To filter mercury from contaminated streams
Erosion breaks	
	To prevent or retard soil erosion on slopes and shore lines
Animal mattresses	
	Race horses and prize cattle
Distillation	
	150 gal of oil; 1500 ft^3 of gas; 40% residue per ton of tires
Sports fields	
	Football fields, tennis courts, track fields, and race tracks
Filler material	
	Thermoset plastics
Soil conditioner	
	Prevents caking and provides aeration
Oil spill clean-ups	
	1 lb of rubber will absorb on the average 4 lb of oil
Pyrolysis	

B. Other Recent Developments

Other special treatments of the particles could also be utilized, such as the production of ion exchange resins. For instance, the Savannah River Ecology Laboratory at Aiken, S. C. is researching the feasibility of using rubber particles to filter mercury from contaminated streams.[28]

Another exciting development is the recovery of metal and insulation from used wire and cable materials. Methods now employed involve the burning or dissolution of the insulation off the metal. These methods are slow, expensive processes and produce either oxidized- and carbonized-coated metal or expensive solvent recovery problems. The resulting metal would often have to be further refined. Cryogenic processing is rapid and clean, and the metal is easily recovered intact.

Electric plugs, sockets, and switches often have high impact phenolics or phthalates as insulators which resist fracture. Cryogenic processing embrittles these materials, and, due to different coefficients of expansion of the insulator and metal, separation is attained more easily.

Laminates of metal and polymer sheets, which usually involve some adhesive, can be separated if there is sufficient embrittlement of the adhesive or polymer. Also, delamination of polymer laminates is readily obtained if one of the polymers becomes brittle. Upon impact, the embrittled polymer fragments and releases itself from the still-flexible portions of the laminate.

By cryogenic processing of coated materials, one can easily remove the polymer coating if the coating, normally a paint formulation, becomes brittle. Adhesion is considerably reduced, especially under impact. A particular application could be the cleaning of hooks and holders for articles being sprayed or dip-coated. Sometimes just dipping into the liquid nitrogen provides enough thermal stress to fracture the coating.

The applications for cryogenic recycling have also been extended to the handling of entire automobiles and large metal pieces. Robert George, an industrialist of Liege, Belgium, freezes the entire automobile for recycling. The freezing makes it possible to use less horsepower for the fragmenting and, even more importantly, aids in separating the nonferrous from the ferrous metals. Freezing does not affect the metals as burning does; therefore, this process produces a high grade of nonferrous metal.

In addition, Stainless Processing Company of Chicago, using cryogenic technology developed by Air Products and Chemical, has processed motors, generators, and electrical equipment.

Perhaps the most recent study is the feasibility of removing the hair from the skin of the hog immediately following its killing. It is reported that the common practice today is to "steam-cook" the hide so that the hair can be more easily removed. In so doing, the hide is conditioned so that it can be used only for gelatin. By chilling to a −68°C (−90°F), the hair can be removed cryogenically, leaving a skin suitable for leather, a much more highly valued product.

The cryogenic process is effective and efficient; the proper market for the separated materials can make it economical. With the priorities of our country put into proper prospective, wide use of the process can become a reality.

It would appear that the cryogenic age for recycling is upon us. It will be a challenge to be a leader in its development.

REFERENCES

1. **Michalski, P. H.,** A Classification of Materials Which Can Be Cryogenically Recycled, M. S. thesis, University of Wisconsin—Madison, 1976, 24.
2. **Michalski, P. H.,** A Classification of Materials Which Can Be Cryogenically Recycled, M. S. thesis, University of Wisconsin—Madison, 1976, 16.
3. **Michalski, P. H.,** A Classification of Materials Which Can Be Cryogenically Recycled, M. S. thesis, University of Wisconsin—Madison, 1976, 15.
4. **Andrews, E. H.,** *Fracture in Polymers,* Oliver & Boyd, London, 1968, 50.
5. **Michalski, P. H.,** A Classification of Materials Which Can Be Cryogenically Recycled, M. S. thesis, University of Wisconsin—Madison, 1976, 27.
6. **Warren, K. A. and Reed, R. P.,** Tensile and Impact Properties of Selected Materials From 20 to 300 K, National Bureau of Standards, Monograph Series No. 63, U.S. Department of Commerce, 1963, 32.
7. **Wessel, E. T.,** Some basic engineering considerations regarding the fracture of metals at cryogenic temperatures, behavior of materials at cryogenic temperatures, Spec. Tech. Publ. 387, American Society for Testing and Materials, Philadelphia, 32.
8. **Warren, K. A. and Reed, R. P.,** Tensile and Impact Properties of Selected Materials From 20 to 300 K, Monogr. Ser. No. 63, National Bureau of Standards, U.S. Department of Commerce, 1963, 36.
9. **Warren, K. A. and Reed, R. P.,** Tensile and Impact Properties of Selected Materials From 20 to 300 K, Monogr. Ser. No. 63, National Bureau of Standards, U.S. Department of Commerce, 1963, 34.
10. **Wessel, E. T.,** Behavior of materials at cryogenic temperatures, Spec. Tech. Publ. 387, American Society for Testing and Materials, Philadelphia, 35.
11. **Michalski, P. H.,** A Classification of Materials Which Can Be Cryogenically Recycled, M. S. thesis, University of Wisconsin—Madison, 1976, 42.
12. **Michalski, P. H.,** A Classification of Materials Which Can Be Cryogenically Recycled, M. S. thesis, University of Wisconsin—Madison, 1976, 50.
13. **Halldin, George,** personal communication 1980.
14. **Van Vlack, L. H.,** *Elements of Materials Science,* Addison Wesley, Reading, Mass., 1964, 190.
15. **Michalski, P. H.,** A Classification of Materials Which Can Be Cryogenically Recycled, M. S. thesis, University of Wisconsin—Madison, 1976, 55.

16. **Michalski, P. H.,** A Classification of Materials Which Can Be Cryogenically Recycled, M. S. thesis, University of Wisconsin—Madison, 1976, 59.
17. **Michalski, P. H.,** A Classification of Materials Which Can Be Cryogenically Recycled, M. S. thesis, University of Wisconsin—Madison, 1976, 57.
18. **Michalski, P. H.,** A Classification of Materials Which Can Be Cryogenically Recycled, M. S. thesis, University of Wisconsin—Madison, 1976, 60.
19. **Andrews, E. H.,** *Fracture in Polymers,* Oliver & Boyd, London, 1963, 53.
20. **Michalski, P. H.,** A Classification of Materials Which Can Be Cryogenically Recycled, M. S. thesis, University of Wisconsin—Madison, 1976, 63.
21. **Kisielewski, R.,** Cryogenic Recycling International, La Cross, Wis., 1977.
22. **Braton, N. R.,** unpublished observation, 1977.
23. Int. Symp. Use of Rubber in Asphalt Pavements, Proceedings, Salt Lake City, Utah, 1971.
24. **Brand, B. G.,** Scrap Rubber Tire Utilization in Road Dressings, EPA-670/2-74-014, Battelle-Columbus Laboratories, Colombus, Ohio, March 1974.
25. **McDonald, C. H.,** The Use of Recycled Rubber in Roads, 58th Ann. Meet. of American Assoc. of State Highway Officials, Phoenix, Ariz., December 1972.
26. **Tagart, R.,** Shredded Tires as an Auxiliary Fuel. Environmental Aspects of Chemical Use in Rubber Processing Operations, Conf. Proc., Publ. No. 560/1-75-002, U.S. Environmental Protection Agency, Washington, D.C., July 1975.
27. **Beckman, J. A. and Laman, J. R.,** Destructive Distillation of Used Tires, Firestone Tire and Rubber Co., Akron, Ohio, 1969.
28. **Tharin, D. W.,** Mercury Removal From Aqueous Process Streams, Contract No. AT107-21-1, du-Pont de Nemours & Company, Savannah River Laboratory, Aiken, S.C., U.S. Atomic Energy Commission.

CHAPTER 3

CRYOGEN PROPERTIES AND PROCESSES

D. Otis

TABLE OF CONTENTS

I. INTRODUCTION

This chapter will provide design information for engineers working perhaps for the first time in cryogenic processing. Much of the discussion applies to cryogens in general, and a representative list with properties is provided in Table 1. Throughout the chapter, however, emphasis is placed upon nitrogen because of the central role it plays in cryogenic processing as it exists today. Where possible, nitrogen data and correlations based upon that data have been presented rather than results based upon data for other substances.

II. PHYSICAL AND THERMODYNAMIC PROPERTIES

The boiling point varies widely for the cryogens shown in Table 1.[1] The range of greatest interest for cryogenic processing is between −101°C (−150°F) and −46°C (−50°F), and as can be seen in the table very few candidates are in that range. Both nitrogen and carbon dioxide are widely used cryogens. Carbon dioxide is more difficult to handle because it is a solid at atmospheric pressure. It can be transported under pressure as a liquid, but at any stage of depressurization, solids can form which can clog valves and lines.

Though nitrogen (−195.8°C [−320.5°F]) is much colder than is required, it has the advantages of low cost, no toxicity hazard, and is inert.

A. Vapor Pressure

Liquid nitrogen boils when its vapor pressure equals 1 atm, and for standard conditions this occurs at a temperature of −195.8°C (−320.5°F), as shown in Figure 1. If liquid boiling at 1 atm is suddenly pressurized to 15 psig (29.7 psia or 2.05 bars), for example, the boiling will immediately cease, and the heat leak (which continues as before) will begin to raise the temperature until it reaches the new boiling point of −189.4°C (−309°F), after which boiling will continue. For each pressure, there is a unique boiling temperature, as indicated in Figure 1. Even if the storage pressure were raised to 350 psia (364.7 psia or 25.2 bars), the boiling temperature would still be quite cold (−153.3°C or −244°F); therefore the problem of boil-off cannot be eliminated to any practical degree by pressurization. It is for this reason that cryogens are stored in low-temperature vessels and with highly effective insulation.

B. Flash Vapor

Reducing the pressure below 1 atm in a Dewar that has been at saturation (i.e., at the boiling condition) results in the immediate formation of flash vapor. For example, if a Dewar at 1 atm is suddenly connected to a vaccum vessel at 10 psia (0.69 bar), flash vapor will form, and a two-phase liquid-vapor mixture will be transferred at a rate determined by the flow resistance of the transfer tube (see Section V). And the temperature of the fluid remaining in the Dewar will drop until it reaches the new boiling temperature of −198.8°C (−326°F) corresponding to 10 psia (Figure 1). Since the liquid temperature can be either raised or lowered by changing the pressure, it cannot be assumed that the liquid is always delivered at precisely the same temperature. Rapid and continuous removal of liquid nitrogen from a Dewar at constant liquid temperature requires external pressurization or external heating. In fact, if the pressure falls below the thermodynamic critical pressure for nitrogen (1.818 psia or 0.125 bar), solid-nitrogen or solid-liquid slush will appear and could cause some trouble in the delivery system. Such conditions, however, would be very unusual.

Table 1
CRYOGENIC AND REFRIGERATING LIQUIDS

Fluid	Boiling point				Density		Latent heat of vaporization		Specific heat, C_p			
	K	°R	°C	°F	Liquid (kg/m³)	Liquid (lb/ft³)	kJ/kg	Btu/lb$_m$	Liquid (kJ/kg·K)	Liquid (Btu/lb$_m$R)	Gas (kJ/kg·K)	Gas (Btu/lb$_m$R)
Air	79	142	−194	−318	875	54.6	205	88.2	1.97	0.470	1.02	0.245
Argon	87	157	−186	−303	1400	87.6	162	69.5	1.14	0.272	0.531	0.127
Carbon dioxide[a]	195	350	−79	−110	1560	97.6	572	246	1.33	0.318	0.795	0.190
Ethane	185	334	−88	−126	548	34.2	488	210	2.51	0.600	1.05	0.250
Fluorine	85	154	−188	−306	1500	94.0	166	71.6	1.55	0.37	0.812	0.194
Helium 3	3.20	5.76	−270	−454	58.9	3.68	8.48	3.65	4.60	1.10		
Helium 4	4.215	7.59	−269	−452	125	7.80	20.7	8.92	4.56	1.09		
Hydrogen	20	36.7	−253	−423	71.0	4.43	446	192	9.79	2.34	6.82	1.63
Methane	111	201	−162	−259	424	26.5	509	219	3.45	0.825	11.9	2.85
Neon	27	48.8	−246	−411	1200	75.2	86.7	37.3	1.84	0.44	1.66	0.398
Nitrogen	77.4	139.2	−196	−320	810	50.6	198	85.3	2.04	0.487	1.17	0.280
Oxygen	90.2	162	−183	−297	1140	36.3	213	183	1.90	0.526	1.08	0.259
Propane	231	416	−42	−44	581	38.3	425	188	2.20	0.619	1.40	0.290
Propylene	225	406	−48	−54	614	92.9	437	71.1	2.59	0.213	1.21	0.363
Refrigerant 12	243	438	−30	−22	1490	95.0	165	64	0.891	0.214	1.52	0.109
Refrigerant 13B1	215	388	−58	−72	1990	124	119	51.1	1.69	0.405	0.456	0.0987
Refrigerant 22	232	419	−41	−41	1410	88.2	234	100.5	1.05	0.252	0.849	0.203

[a] Sublimes.

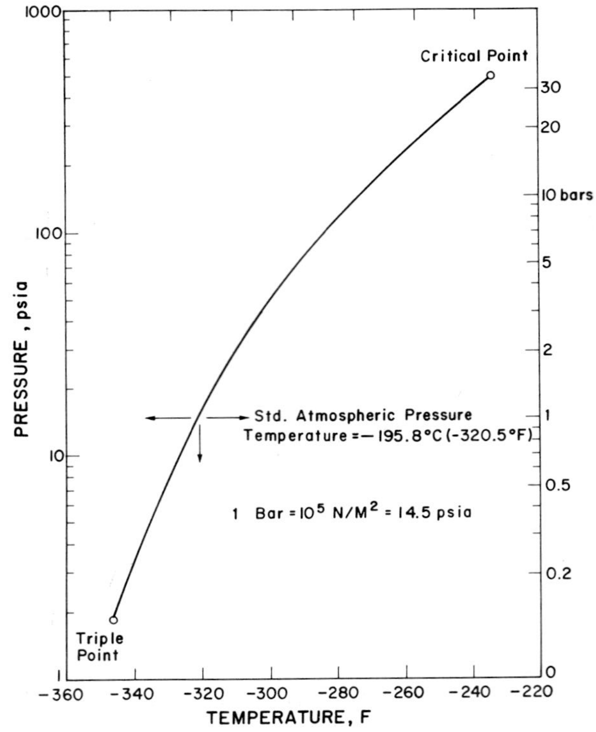

FIGURE 1. Vapor pressure for nitrogen liquid and gas in equilibrium as a function of temperature. The vapor pressure is one standard atmosphere at a temperature of −195.8°C (−320.5°F), the temperature normally associated with boiling nitrogen; however, this temperature depends upon the pressure.

C. Thermodynamic Properties For Nitrogen

Figure 2* presents the thermodynamic properties for nitrogen in the gas, liquid, and the gas-liquid regions. One important quantity that can be determined from Figure 2 is the amount of energy that can be absorbed by the liquid as it boils and heats up to the ambient temperature. This process is traced out on the diagram for a pressure of one standard atmosphere. The state of the saturated liquid (liquid just ready to boil) is located at point A. Boiling occurs from A to B with state B referred to as saturated vapor (gas just ready to condense).

An unfortunate property of nitrogen and most other cryogens is their low latent heat of vaporization. This is the amount of heat required to change a unit mass from the liquid to the gas state (from A to B). For nitrogen this equals 200 J/g (85.8 Btu/lb) at 1 atm as compared with 2300 J/g (1000 Btu/lb) for water. With cryogens stored at or just below the boiling point, it takes only a small heat leak to produce vapor by boiling, and flowing cryogens are usually two-phase mixtures of the liquid and gas phases.

The saturated vapor (point B) is still at −195.8°C (−320.5°F), but as it continues to warm up it follows the 1 atm constant pressure line to point C. During this process

* Figure 2 is obtainable from the NBS Cryogenic Engineering Laboratory,[2] and other charts and tabular data[3] are available.[4] A computer program package has been developed by Pew, Hendricks, and Simoneau[5] for nitrogen thermodynamics and transport properties, and an equation of state for the liquid and gas has been formulated for use with computer calculations by Jacobson, Stewart, and Myers.[6]

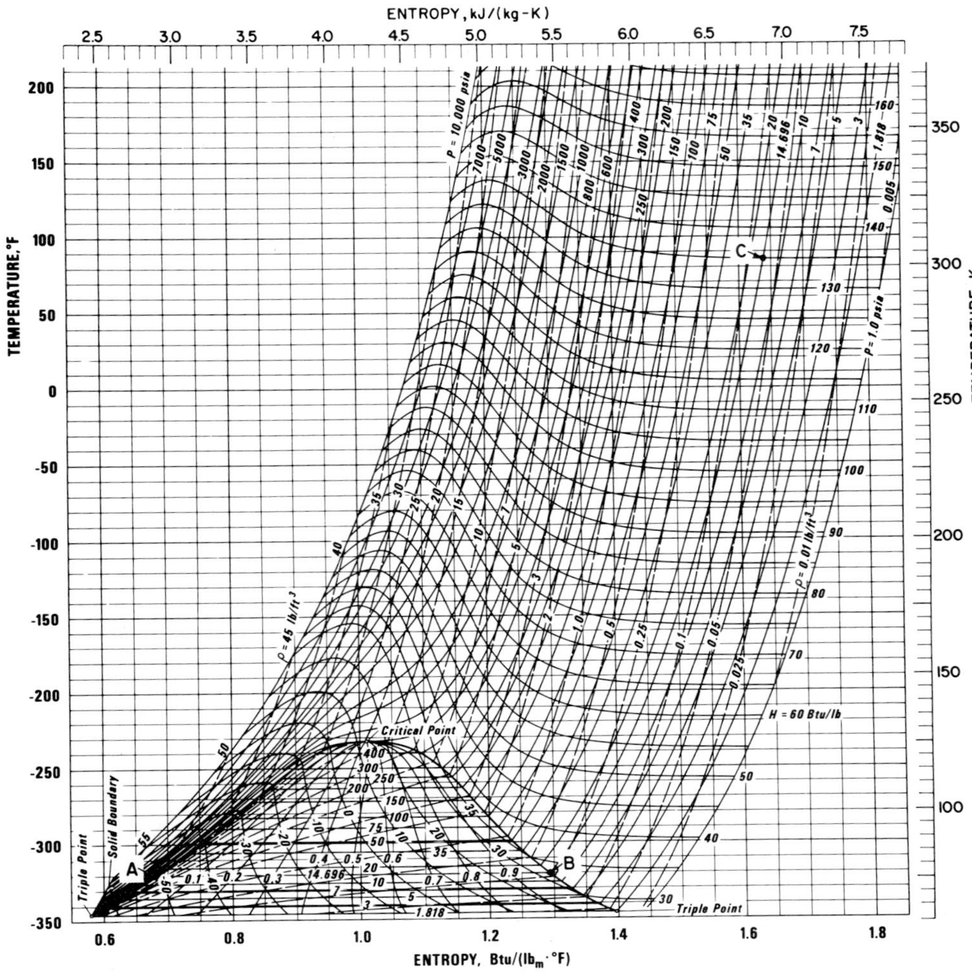

FIGURE 2. Temperature-entropy chart for nitrogen. (Modified from National Bureau of Standards, chart 2027W (4-1-77), NBS-IBS Boulder, Colo. The 22 × 33 in. chart is available for $1.00 from NBS.)

from B to C, it is said that the nitrogen is absorbing "sensible heat" in contrast to the "latent heat" referred to previously for process A-B. The sensible heat for nitrogen warming to 21°C (70°F) is 228.8 J/g (98.4 Btu/lb), and the total energy absorbed in the entire process is 428.3 J/g (184.2 Btu/lb). This is the maximum "total refrigeration effect" for liquid nitrogen, and, as indicated in Figure 3, it is subdivided into a "latent refrigeration effect" and a "sensible refrigeration effect". In some cryogenic processing applications, only the latent heat will be realized, as the gas may escape the process before it warms. An attempt should be made to design the process so as to utilize the sensible heat also, by not permitting the gas to leave the process until it is warmed to near ambient temperature.

III. INSULATION AND DEWARS

The thermal properties of cryogenic insulation are discussed by Tien and Cummington,[7] and some typical values are presented in Table 2. As shown in Figure 4, a vacuum-jacketed thermos bottle will hold liquid nitrogen satisfactorily for several hours

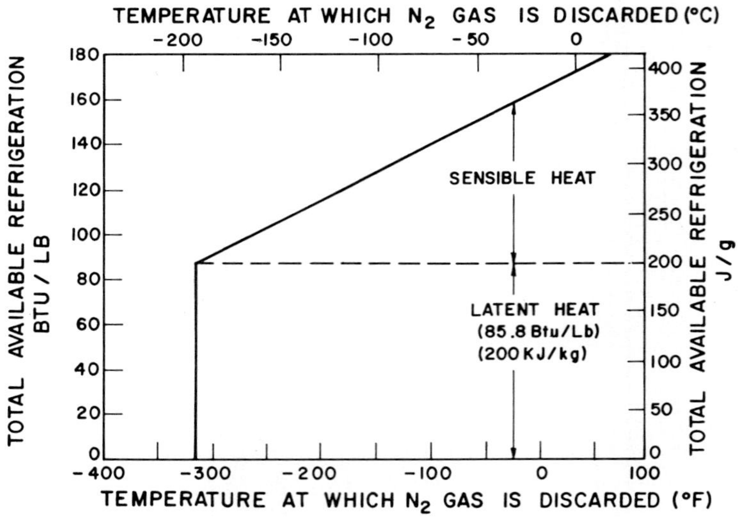

FIGURE 3. Refrigeration obtainable from liquid nitrogen as a function of the temperature at which the nitrogen gas is discarded from the process. (From Union Carbide Corporation, N_2, publ. F-4218, with permission.)

(refer to Chapter 7 for a full discussion of cryogenic storage vessels). A tight-fitting stopper should not be used without a pressure relief valve.

Since cryoprocessing usually involves fairly large quantities of materials for short periods of time, insulation is not as critical as it would be for aerospace applications. Plastic foams provide a satisfactory and economical solution to cryoprocessing insulation requirements. Commercial Dewars are available with either foam or vacuum insulation, and specially shaped containers can be constructed from urethane foam sheets (1- to 2-in. thickness is adequate in most cases).

IV. HEAT TRANSFER

Chilling the process materials to their embrittlement temperature can be accomplished by direct immersion in liquid nitrogen, by liquid spray, or by contact only with the gas. In either case, the refrigeration involves heat transfer, and the primary questions to be answered are (1) What time is required? (2) What quantity of liquid nitrogen is consumed?

If a solid body at room temperature is immersed in liquid nitrogen, the temperature drops dramatically, as shown in Figure 5. During the first phase of cooling, the object is blanketed by the nitrogen gas produced from the boiling, and this "blanket" produces an insulating effect. As the surface temperature drops to about −160°C (−256°F), a sudden transformation occurs in the boiling mechanism as the blanket film becomes unstable and breaks up. The heat transfer has experienced a transition from "film boiling" to "nucleate boiling". This transition results in a more than tenfold increase in the heat transfer rate, and within just a few seconds the body reaches equilibrium, as shown in Figure 5. Points A-B-C-D trace out the process in both Figures 5 and 6. For cooldowns, most of the time is spent in the film boiling region between A and B, and about 90% of the refrigeration has been accomplished upon arriving at B. A finger quickly thrust in and out of liquid nitrogen will escape frostbite because of the protective film blanket of film boiling. A drop of liquid nitrogen scurries around the floor for a time, riding on a protective film of nitrogen gas until tran-

Table 2
CRYOGENIC INSULATION PROPERTIES

Descriptive name	Approximate density		Approximate specific heat		Range of mean conductivities		Interspace pressure
	lbm/ft³	kg/m³	Btu/lbm °F	kJ/kg K	Btu/hr ft °F	mW/space m K	(mm Hg[a])
Multilayer	5	80	.22	0.92	.000023—.00012	0.04—0.2	10^{-4}
Opacified powder	7	110	.23	0.96	.00015—.0004	0.26—0.7	10^{-4}
Evacuated powder	6	100	.25	1.05	.00057—.00115	1.0—2.0	10^{-4}
Vacuum flask	—	—	—	—	.0029	5.0	10^{-6}
Gas-filled powder	6	100	.25	1.05	.001—.004	1.7—7.0	760
Expanded foam	2	30	0.4	1.67	.0029—.020	5.0—35	760
Fiber blanket	8	130	0.5	2.09	.02—.026	35—45	760

[a] For N/m^2 multiply by 133.32.

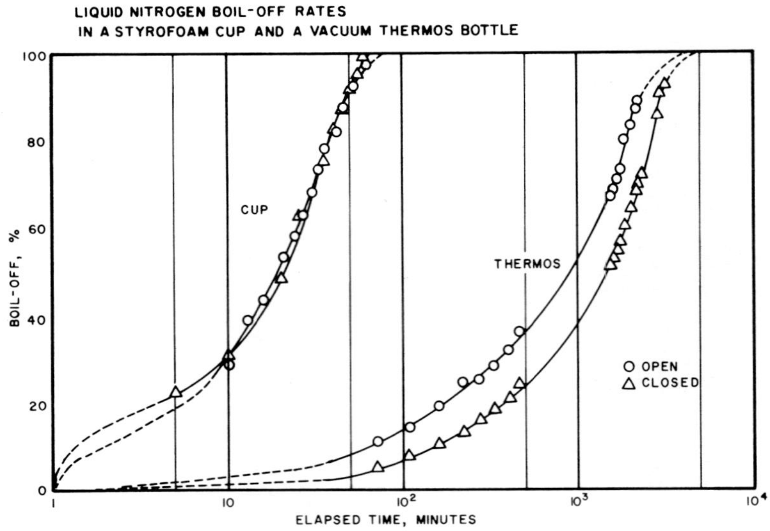

FIGURE 4. Time required for evaporation of liquid nitrogen from a 1 ℓ vacuum thermos bottle and from a styrofoam cup (uncovered) both initially full.

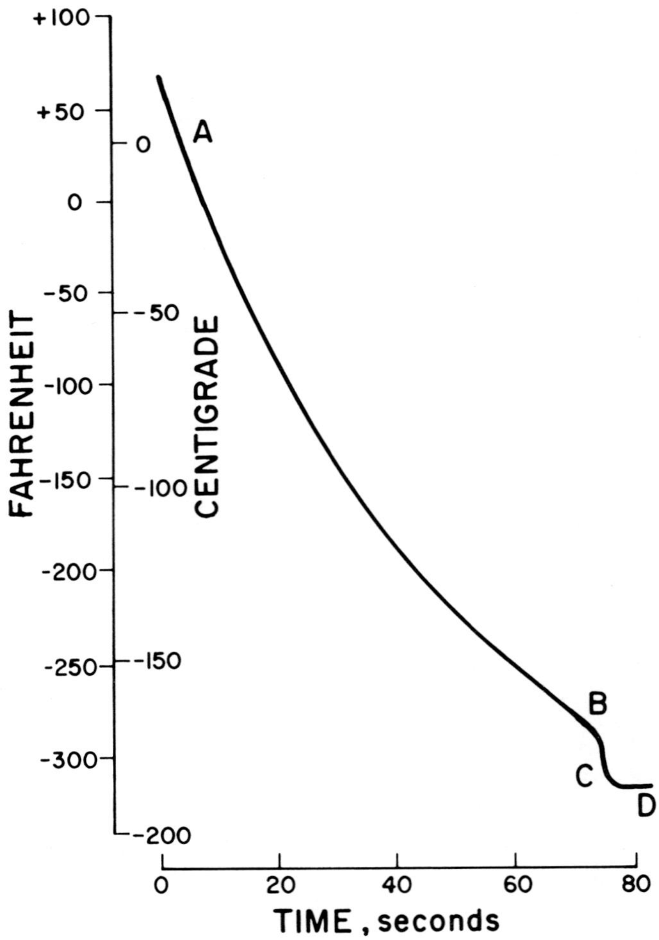

FIGURE 5. Cooldown curve for a brass cylinder (1.2 cm × 1.2 cm diameter). Immersed in liquid nitrogen.

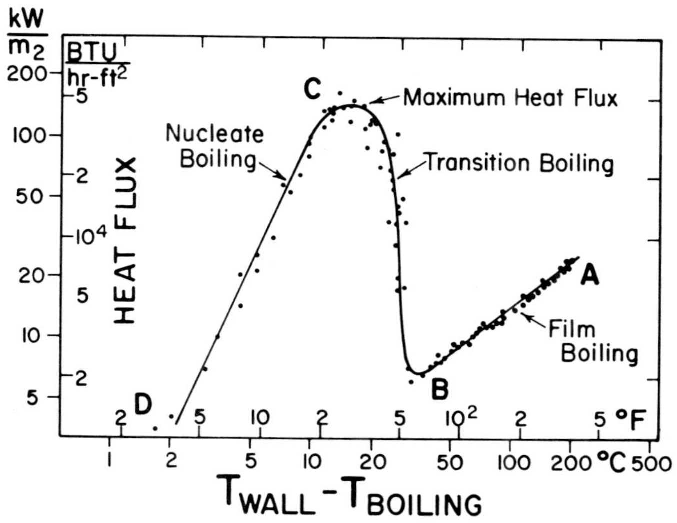

FIGURE 6. Boiling heat transfer curve for liquid nitrogen.

sition to nucleate boiling consumes it with a hiss. The transition from B to C is accompanied by the noise of bubble agitation and occurs at the end of the cooldown period.

Most of the gas produced during cooldown rises through the liquid as bubbles and leaves the surface at a temperature near that of the liquid. This means that only the latent heat has been utilized, and the cold gas leaving the surface is still capable of absorbing a sensible heat of 228.8 J/g (98.4 Btu/lb). This sensible refrigeration capacity should not be wasted.

The C-D portion of the boiling curve (Figure 6) is of interest when heating surfaces which are already cooled down to −195.8°C (−320.5°F). If, for example, an electric wire is heated which is submerged in liquid nitrogen, the heat transfer rate could be increased to almost 150 kW/m^2 (47,550 Btu/ft^{2}hr) with a temperature rise of only about 15°C (27°F). A small additional increase in heat rate, however, requires a sudden jump in wire temperature to well over 550°C (1022°F), a point off the figure. Point C is often referred to as ''burnout'', and wires burn out (or at least glow red to white hot) when this condition is exceeded in liquid nitrogen. This maximum or ''critical heat flux'' represents the upper limit on practical heat rates that can be attained. Operating near this condition is apt to be unstable, especially with small wires which are inclined to develop localized hot spots. This would produce a local increase in the electrical resistance which at constant current means even greater heat flux and consequent burnout. Power transients can reduce the burnout heat flux to as little as 40% of the steady state value.[9]

Whenever the cold surfaces in contact with liquid nitrogen are heated (purposely or by heat leaks), there is usually boiling. For a well insulated Dewar, the heat flux is so low that bubbles may not appear on the walls, but they will form at the upper surface when exposed to warm air. In transfer lines, the cooldown heat load initially boils away all of the liquid, and only gas is delivered until the wall temperatures have dropped substantially. It is best to use a reduced flow during cooldown of transfer lines (if time permits) to make maximum use of the sensible refrigeration effect. Even after cooldown, some boiling occurs in the transfer line due to heat leak and pressure drop (see next section), and the flow contains both the gas and liquid phases (two-phase flow).

Cooldown liquid nitrogen requirements can be divided into two categories. For De-

Table 3
ENTHALPY CHANGE REQUIRED TO REDUCE TEMPERATURE FROM 25°C (77°F)

	Final temperature		Refrigeration required	
Substance	°C	°F	kJ/Kg	Btu/Lbn
Tire rubber	−25	−13	84.9	36.5
	−50	−58	119.8	51.5
	−75	−103	145.3	62.5
	−125	−193	177.9	76.5
	−195.8	−320.5	210.5	90.5
ABS	−75	−103	125.6	54.0
	−195.8	−320.5	232.5	100
HYCAR rubber	−75	−103	118.5	51
	−195.8	−320.5	241.8	104
Epoxy resin	−25	−13	39.5	17
	−75	−103	79.1	34
	−195.8	−320.5	144.2	62
Nylon	−25	−13	65.1	28
	−75	−103	130.2	56
	−195.8	−320.5	220.1	95
PVC	−25	−13	51.2	22
	−75	−103	102.3	44
	−195.8	−320.5	181.4	78
Crushed chili peppers	−25	−13	62.8	27
	−75	−103	123.3	53
	−195.8	−320.5	200.0	86

wars and plumbing that will contain the liquid nitrogen, their temperature must be reduced to −195.8°C (−320.5°F). On the other hand, for process materials, their temperature must be reduced only to their required processing temperature (usually their embrittlement temperature). Table 3 provides some approximate values for the refrigeration required to attain various temperatures.[10] The embrittlement temperature is not specified since its value depends upon the type of test and other factors.[11] In most cases, embrittlement temperatures and refrigeration requirement must be established by measurement in each case. Jacobs[12] has presented methods for calculating the liquid requirements for cooldown, and calculations are given for 18-8 stainless steel, copper, and aluminum.

Cooldown times vary widely depending upon the specimen size and thermal properties. Some representative cooldown times for complete immersion are listed in Table 4 for cooling to −195.8°C (−320.5°F). Shorter times would be required to reach embrittlement temperature. For large objects, cooldown to embrittlement temperature is complicated by the nonuniformity of the temperature distributions established in the solid specimens. Figure 7 shows the temperature history for three thermocouples placed at various depths of a rubber tire casing. Upon immersion in liquid nitrogen, the surface thermocouple dropped rapidly below −70°C (−94°F) while the interior thermocouples responded more slowly. After 30 sec, the specimen was removed from the liquid nitrogen and placed in a styrofoam box to "soak" for 45 sec, and an equilibrium temperature of −56°C (−68°F) was attained. Figure 7 illustrates the need for immersion followed by a soaking period to attain a uniform temperature.

Although not a property of the cryogen, the warmup of cryogenically processed materials is a practical concern. The milled product is usually almost as cold after milling as before, and frost forms on surfaces exposed to the atmosphere. Warmup

Table 4
COOLDOWN TIMES FOR MATERIALS SUBMERGED IN LIQUID NITROGEN

	Final temperature			
	−50	−100	−150	−195.8°C
Sample	−58	−148	−238	−320.5°F
	Time in seconds			
Aluminum	11	32	57	75
Copper	35	70	113	128
Graphite	8	16	23	28
Masonite	—	—	—	70
Plexiglass	—	—	—	150
Rubber	—	—	—	140
Stainless steel (1020)	—	35	85	112

Note: All samples are rectangular blocks 2.54 cm × 2.54 cm × 1.27 cm (1 in. × 1 in. × ½ in.) instrumented with a single thermocouple at the center. Initial temperature = 25°C.

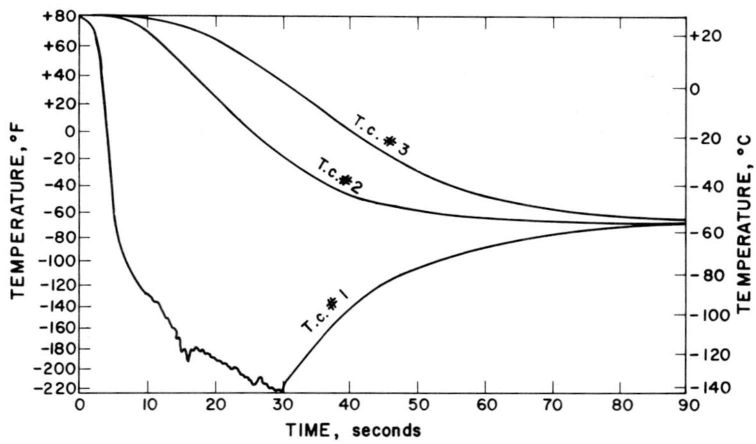

FIGURE 7. Temperature history in a tire casing which is immersed in liquid nitrogen for 30 sec, and then removed from the liquid. Thermocouple No. 1 was on the surface, and No. 2 and No. 3 were at interior locations.

time varies widely depending upon the size and thermal properties of the material. A piece of automobile tire casing 1.27 cm × 10.16 cm × 10.16 cm (½ in. × 4 in. × 4 in.) warms up in about 15 min. A pile of milled product could retain the cold for much longer (hours), which is undesirable from the standpoint of further handling, frost, and dew formation, or icing if exposed to a light rain. It would seem most advantageous to recover the cold and reject the processed material at a temperature above freezing. This not only avoids the aforementioned problems, but may enable some recovery of the cold and thereby reduce liquid nitrogen consumption.

In this regard, the input materials can be placed in direct contact with the cold processed materials. A direct contact heat exchange utilizing conveyor systems or fluidized

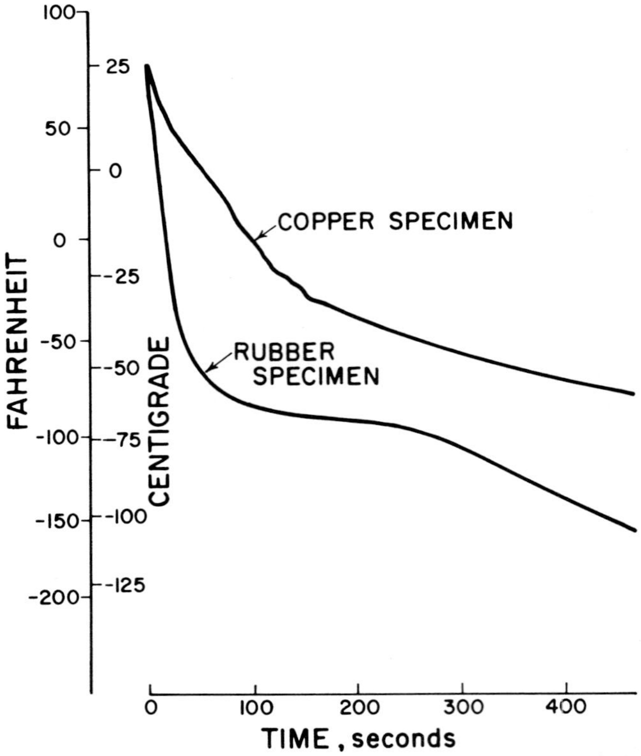

FIGURE 8. Cooldown curve for copper and rubber blocks (2.54 cm × 2.54 cm × 1.27 cm) when submerged in 35 mesh milled rubber initially at liquid nitrogen temperature.

bed are suggested. Solid copper and rubber blocks (2.54 cm × 2.54 cm × 1.27 cm) are surrounded by cold milled rubber (35 mesh), and the temperature histories of the blocks are shown in Figure 8. Several minutes were required to attain equilibration even for this small sample, and some form of agitation would be required to reduce the time for energy transfer to the processed rubber.

V. PRESSURE DROP AND COMPRESSIBLE FLOW PHENOMENA

Cryogens behave as other fluids with regard to pressure drop in pipes and choking phenomena. Friction factors can be obtained from the well-known Moody chart. Nozzle flows can be calculated using Bernoulli's equation for the liquid, and isentropic flow tables for the gas when those phases occur singly in the piping system. However, this is seldom the case; the gas phase invariably appears with the liquid due to flashing (caused by pressure drop) and boiling (caused by heat leaks). The resulting two-phase mixture can appear in many forms, as suggested by these names used for the many observed flow patterns: bubble, slug, plug, annular, stratified, spray, and mist. The flow pattern depends upon the pipe orientation, fluid velocity (and the two phases may have different velocities), and the fluid properties.

An interesting and potentially dangerous example of two-phase flow occurs when an open-ended tube is partially immersed in liquid nitrogen. The liquid boils inside the tube, and the two-phase mixture rises in the tube due to its lower density. This results in what can be a rather forceful "cryogenic fountain" operating similar to a percolator coffee pot. A 0.6-m (2-ft) length of ¼ in. I.D. rubber vacuum hose can produce a

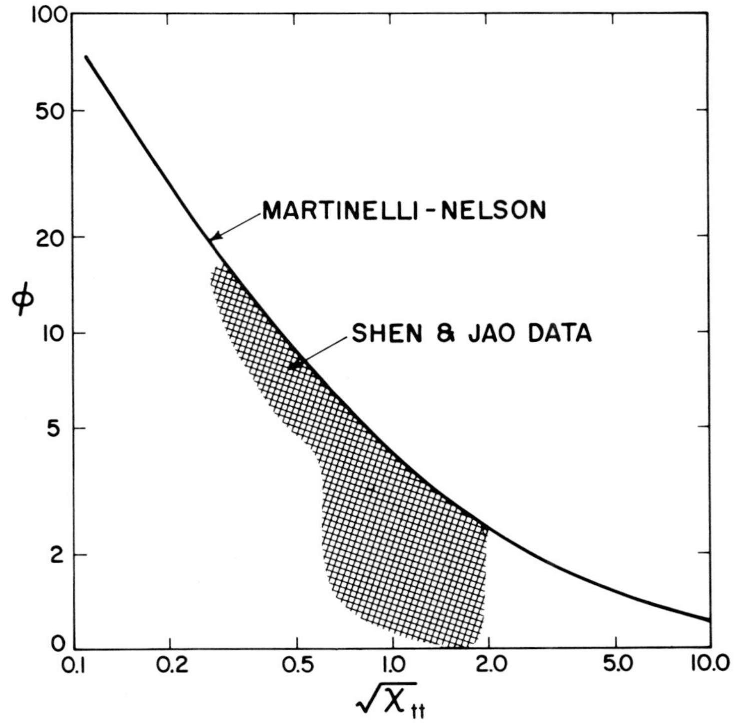

FIGURE 9. The square root of the Lockhart-Martinelli two-phase friction multiplier.

fountain over 0.6 m high when partially immersed in a quart thermos filled with liquid nitrogen.

No attempt will be made here to present details of the several methods and correlations which can be found in specialized works (Jacobs,[12] Collins,[13] Butterworth and Hewitt,[14] and Frost[15]). As the quality of the flow increases from zero (all liquid) to 1.0 (all gas), the frictional pressure drop increases dramatically. The "two-phase frictional multiplier" ϕ^2 is the factor by which the pressure drop is increased over what it would be if only the liquid phase flowed. The Lockhart-Martinelli correlation[16] for ϕ is plotted in Figure 9. Shen and Jao[17] have obtained data to show that the Lockhart-Martinelli correlation provides reasonable estimates for liquid nitrogen. The abscissa of Figure 9, χ_{tt} is obtained from:

$$\chi_{tt} = \frac{1 - x}{x} \left(\frac{\rho_g}{\rho_f}\right)^{0.571} \left(\frac{\mu_f}{\mu_g}\right)^{0.145} \qquad (1)$$

where x = quality, ϱ = density, μ = viscosity, and the subscripts f and g refer to the liquid and gas phases, respectively. A plot of this equation, as presented by Smith,[18] is shown in Figure 10 for liquid nitrogen. The pressure drop expected, if all the flow were the liquid phase, is calculated and then is multiplied by the factor ϕ^2 to estimate the two-phase frictional pressure drop. Note that this is accurate only for pressure drops that are small as compared with the average pressure level in the pipe, and there is also another component to the pressure drop caused by the acceleration of the fluid as the quality increases.

The important result to be noted here is that pressure drop increases dramatically

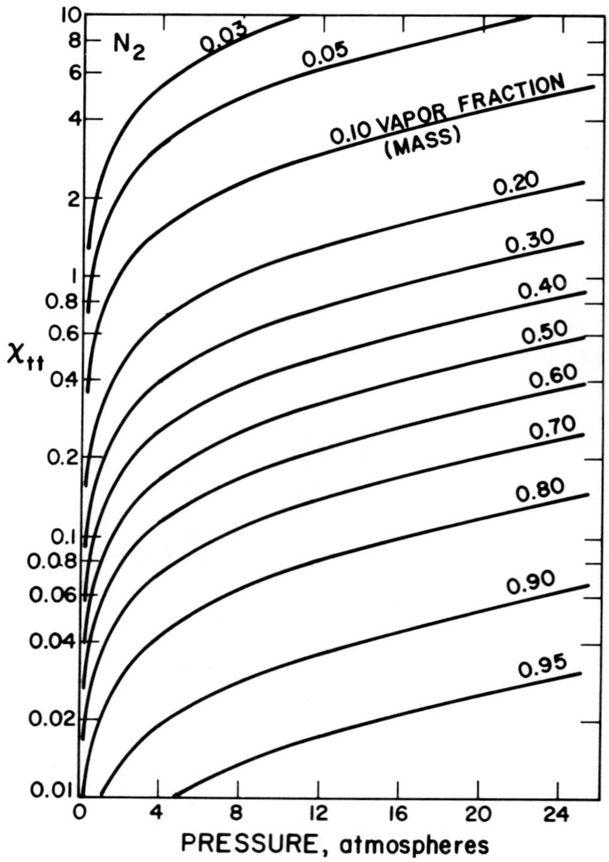

FIGURE 10. Martinelli-Nelson correlating term x_{tt} as presented by Smith for nitrogen. (With permission.)

as vapor is formed. The pressure drop "feeds on itself" by producing more flash vapor. For this reason, it is important to vent the vapor in long tubes at regular intervals if large pressure drops and the associated flash vapor generation are to be avoided.

The flash vapor produced when the saturated liquid pressure is decreased at constant enthalpy (throttling across a valve, or frictional pressure drop) is shown in Figure 11, as prepared by Union Carbide Corporation.[19] As illustrated by Process A-B, throttling liquid from 50 psig (point A) to 10 psig (point B) produces 11% by weight of vapor ($x = 0.11$). The vapor density is much less than the liquid density, and a quality of 0.11 results in a void fraction of about 90%. Therefore, even at low qualities, the effluent appears to be mostly gas.

Choking (attainment of the speed of sound) can occur at fairly low velocities in two-phase flows because the speed of sound for the mixture is usually much lower than for either the liquid or gas phase. In a tube, choking occurs at the tube exit, and the flow through the tube cannot be further increased by lowering the downstream pressure. Tubing must be sized so that the mass velocities (defined as the product of density and velocity) are well below the critical values. A summary of the experimental data of Brennan, Edmonds, and Smith[20] is given in Table 5 as a guideline for design purposes.

Flow instabilities are certain to be encountered in cryogenic systems. Water hammer, column separation, and self-excited, thermally induced oscillations can be noisy and destructive if excessive. The problems are not unlike those experienced in steam sys-

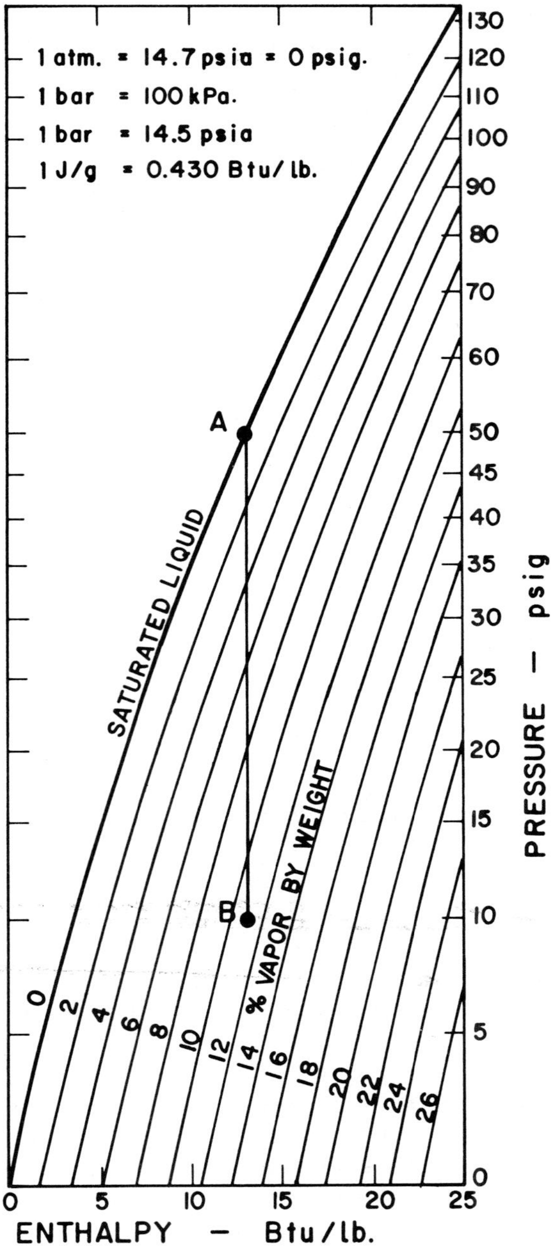

FIGURE 11. Pressure-enthalpy diagram for nitrogen. The flash vapor produced by throttling (an adiabatic pressure drop) can be determined by following a vertical line as illustrated. (From Union Carbide Corporation, N_2, Publication F-4218, Linde Division, New York, N.Y., with permission.)

tems. Any sudden valve closure in a liquid line will result in column separation (creation of a void downstream of the valve), and the return impact upon collapse of the void can produce high peak pressures. Fortunately, cryogenic system pressures are modest, and in most cases the instabilities only constitute a nuisance. To avoid cavitation, pumps must be of special design with low net positive suction head (NPSH) requirements and should be located low in the system.

Table 5
MASS VELOCITY ATTAINABLE AT THE EXIT OF A PIPE IN CHOKED TWO-PHASE FLOW OF NITROGEN

Quality[a] (%)	Exit pressure		Mass velocity[b]	
	Bars[c]	psia	(kg/m² sec)	Lbm/ft² sec
2.5—5.0	2	29.0	10,000	2044
5.0—10.0	1	14.5	3,500	715
	2	29.0	8,000	1635
	4	58.0	12,000	2453
10.0—15.0	1	14.5	2,900	593
	2	29.0	5,000	1022
	4	58.0	10,000	2044
15.0—20.0	1	14.5	2,800	572
	2	29.0	4,800	981
	4	58.0	8,000	1635
20.0—25.0	1	14.5	2,300	470
	2	29.0	4,500	920

Note: These values approximately represent the data of Brennan, Edmonds, and Smith.[20]

[a] Quality is ratio of vapor weight to liquid weight.
[b] Mass velocity is the mass flow rate per unit area.
[c] 1 bar = 100 kPa = 14.5 psia.

VI. FRICTION AND WEAR

The effect of a cold environment on friction and wear must be considered in the design of bearings and seals. The coefficient of friction of process materials determines how they will behave during handling such as fluidization properties, and angle of repose when piled. Wisander and colleagues at NASA—Lewis began in 1956 to measure friction and wear of metals and plastics commonly used in cryogenic systems.[21] In their apparatus a 304 stainless steel disk 6.35 cm × 1.27 cm (2½ in. diameter and ½ in. thick) rotated at constant speed submerged in the liquid. A rider with a 0.8 cm (3/16 in.) diameter hemispherical tip was pressed against the rotating disk with a constant load of 1 kg (2.2 lb). Various materials were used for the hemispherical tip. Friction force was determined by measuring the tangential force on the rider, and wear was measured in different ways including weighing the specimens and measuring the axial motion of the rider with a dial indicator. Some of the NASA data is summarized in Table 6.

Cryogens make poor lubricants. They have low viscosity, and easily vaporize leaving a dry contact. The stainless steels commonly used at low temperatures have poor sliding performance with a high coefficient of friction (typically 0.8), and they show rapid wear with a tendency for surface-welding and metal transfer. Teflon with 10% graphite fill has extremely low friction and wear in liquid nitrogen. It appears to be a fairly general result that friction and wear are reduced for virtually all nonmetals by the reduction in temperature to −195.8°C (−320.5°F). This is certainly very obvious for a rubber eraser. At room temperature it has high friction and wear, but at temperatures below −73.3°C (−100°F) it becomes very slippery and will not erase.

Table 6
FRICTION AND WEAR FOR SLIDING CONTACT BETWEEN SMOOTH SURFACES SUBMERGED IN LIQUID NITROGEN

Surface materials	Coeff of friction	Wear (relative scale)
Metals		
304 on 304	.78	6
1020 on 52100	.6	1
Ag on 304	.8	0.56
Ag + M_oS_2 on 304	.33	0.6
Nonmetals on 304 stainless steel		
Carbons (phenol)	0.4—0.6	12—63
Carbons (haloid)	0.58	1
Kel-F®	0.45	40
Teflon®	0.25	0.3
Teflon® + graphite	0.1	0.01

Note: Compiled from Wisander, D. W. and Johnson, R. L., *Adv. Cryog. Eng.*, 4, 71, 1960.

VII. SUMMARY

Cryogens are like other substances in many of their properties, but certain features of their behavior are unusual and result in the unexpected. This chapter highlights some of these features and provides a reference base for further exploration into the subject.

REFERENCES

1. **Bolz, R. E. and Tuve, G. L.**, Eds., *Handbook of Tables for Applied Engineering Science*, 2nd ed., CRC Press, Cleveland, 1976.
2. National Bureau of Standards, Cryogenics Division, Chart 2027W (4-1-77), NBS-IBS, Boulder, Colo.
3. **Jacobsen, R. T., Stewart, R. B., McCarty, R. D., and Hanley, H. J. M.**, Thermophysical properties of nitrogen from the fusion line to 3500R (1944 K) for pressures to 150,000 psia (10342×10^5 N/m^2), Tech. Note 648, National Bureau of Standards, Boulder, Colo., December 1973.
4. **Frizen, D. J. and Mendenhall, J. R.**, Publications and services of the cryogenics division, 1953—1977, Tech. Note 1005, National Bureau of Standards, April 1978.
5. **Pew, K. J., Hendricks, R. C., and Simonean, R. J.**, NTWO — a nitrogen-properties package, *Adv. Cryog. Eng.*, 16, 78, 1970.
6. **Jacobsen, R. T., Stewart, R. B., and Myers, A. F.**, An equation of state for oxygen and nitrogen, *Adv. Cryog. Eng.*, 18, 248, 1972.
7. **Tien, C. L. and Cummington, G. R.**, Cryogenic insulation heat transfer, *Adv. in Heat Transfer*, 9, 349, 1973.
8. **Merte, H. and Clark, J. A.**, Boiling heat-transfer data for liquid nitrogen at standard and near-zero gravity, *Adv. Cryog. Eng.*, 7, 546, 1962.
9. **Sinha, D. N., Brodie, L. C., Semura, J. S., and Young, F. M.**, Premature transition to stable film boiling initiated by power transients in liquid nitrogen, *Cryogenics*, 225, April, 1979.
10. **Ork, J. C.**, Liquid Carbonic Corporation, Technical Sales Dept., private communication.
11. **Braton, N. R. and Michalski, P. H.**, Cryogenic shredding and resource recovery, Proc. 23rd Ontario Industrial Waste Conference, June 13—16, 1976, Toronto, Ontario, 358.

12. **Jacobs, R. B.,** Liquid requirements for the cool-down of cryogenic equipment, *Adv. Cryog. Eng.,* 8, 529, 1963.
13. **Collins, J. G.,** *Convective Boiling and Condensation,* McGraw Hill, New York, 1972.
14. **Butterworth, D. and Hewitt, G. F.,** Eds., *Two-Phase Flow and Heat Transfer,* Oxford University Press, New York, 1977.
15. **Frost, W.,** *Heat Transfer at Low Temperatures,* Frost, W., Ed., Plenum Press, New York, N.Y., 1975, chap. 11.
16. **Lockhart, R. W. and Martinelli, R. C.,** Proposed correlation of data for isothermal two-phase, two-component flow in pipes, *Chem. Eng. Prog.,* 45, 39, 1949.
17. **Shen, P. S. and Jao, Y. W.,** Pressure drop of two-phase flow in a pipeline with longitudinal variations in heat flux, *Adv. Cryog. Eng.,* 15, 378, 1969.
18. **Smith, R. V.,** Choking two-phase flow literature summary and idealized design solutions for hydrogen, nitrogen, oxygen, and refrigerants 12 and 11, Tech. Note 179, National Bureau of Standards, Boulder, Colo., August 1963.
19. Union Carbide Corporation, N_2, Publ. F-4218, Union Carbide Corp., Linde Div., New York.
20. **Brennan, J. A., Edmonds, D. K., and Smith, R. V.,** Two-phase (liquid-vapor), mass limiting flow with hydrogen and nitrogen, Tech. Note 359, National Bureau of Standards, Boulder, Colo., 1968.
21. **Wisander, D. W. and Johnson, R. L.,** Wear and friction in liquid nitrogen with austenitic stainless steel having various surface coatings, *Adv. Cryog. Eng.,* 4, 71, 1960.

Chapter 4

AIR-SEPARATION PLANT SYSTEM TO PRODUCE CRYOGENS

J. M. Venetucci

TABLE OF CONTENTS

I. INTRODUCTION TO CRYOGENICS

A Dutch physicist, Kamerlingh Onnes, required a word to refer to gases that could be refrigerated and become low-temperature boiling liquids which he and others could produce in his laboratory studies. He combined two Greek words to fit his needs: namely, κρῦος (meaning cold) and −γεύες meaning a substance that generates. Hence, cold generating or the word "cryogen" was formed to describe a low-temperature boiling liquid.

The field of studies, development, and applied use of a cryogen is, therefore, easily transposed into "cryogenics" and embraces the science that deals with the equipment, processes, and products involving very low-temperature refrigeration. Applied cryogenics requires the proper use of complicated principles which are distinctively different from other fields and therefore has emerged as a science in itself.

The field of cryogenics can be confined to a predetermined temperature range which can be arbitrarily established by individual choice. However, a normally accepted range can easily be established to indicate temperatures at which special methods of refrigeration and insulation are required, thereby excluding conventional refrigeration equipment. Cryogenic engineers have generally agreed that for temperatures which are lower than −128.89°C (−200°F), the term cryogenics can be applied.

Small quantities of cryogens can be produced in laboratories and used for research studies. However, when large volumes of cryogens are required to supply industrial, medical, or scientific programs, it becomes necessary to produce the cryogen in large plants specifically designed to produce tonnage quantities on a daily basis.

II. AIR-SEPARATION PRODUCTION CAPACITY

The most widely accepted method used to produce "tonnage" amounts of cryogens is by the fractional distillation of liquefied air. This economical process is commonly referred to as "air separation".

Large plant facilities for tonnage production of air-separation products began to become a reality in the U.S. sometime after the end of World War II. Prior to that time, relatively small-sized gaseous plants were supplying the nation's demand for gaseous air products such as oxygen and nitrogen. A typical air-separation plant and the operating control room are shown in Figures 1 and 2.

There was considerable change in the compressed gas industry during the period of time when the new use of tonnage oxygen for basic oxygen furnace steelmaking was being developed. The demand for tremendous amounts of oxygen at extremely high flow rates for steel production created the need for large oxygen tonnage plants. The production capacity of air-separation plants can range from 90 to 1890 metric tons (100 to 2100 tons)/day. To meet the need of the steel industry, air-separation plants began to produce, store, and transport oxygen and the by-product nitrogen in the liquid form. Tonnage plants quickly became capable of supplying the oxygen needs of the steel industry and the by-product nitrogen from the plants was sometimes vented back to the atmosphere. Soon, however, efforts were directed towards using nitrogen. Applications in which liquid nitrogen could be economically used were developed, and the field of applied cryogenics emerged dramatically. Liquid nitrogen is now being used for food freezing, grinding, metal shrinkage, oil well stimulation, low temperature studies, cryosurgery, and artificial insemination.

Tonnage air-separation plants are classified as merchant or captive plants. Captive plants are generally built and operated on the basis of supplying all of the produced product to only one consumer. For example, a large steel company may agree to pur-

FIGURE 1. Typical air separation plant. (Courtesy of Liquid Carbonic Corporation, Chicago.)

FIGURE 2. Typical control room for an air separation plant. (Courtesy of Liquid Carbonic Corporation, Chicago.)

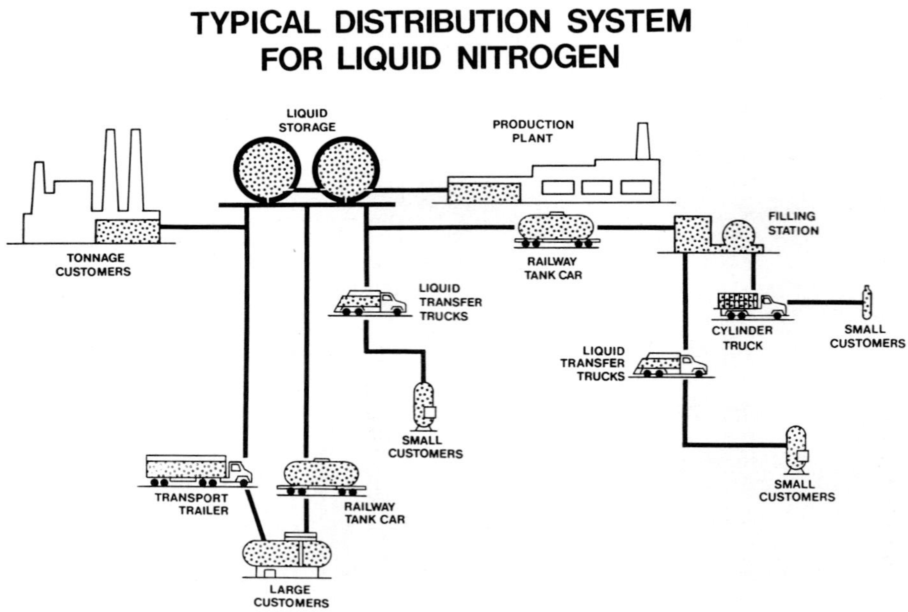

FIGURE 3. Merchant plant product distribution chart. (Courtesy of Liquid Carbonic Corporation, Chicago.)

chase all of the product it requires from an air-separation plant located on or near its property. This type of a tonnage plant would be referred to as a "captive plant" with few if any, products available to other consumers. Obviously, captive plants operate on long-term agreements, with fixed monthly charges, and in turn supply products to the consumer at attractive market prices.

In contrast, "merchant" plants sell the cryogenic product to numerous individual consumers within the area of the plant. The cost to operate a merchant plant is generally higher because of the need for such items as costly mobile distribution equipment, higher electrical power, and additional administrative overhead (Figure 3).

With the demand for cryogenic products such as oxygen and nitrogen growing at an expected rate of 10 to 12% per year, as shown in Figure 4, it has been necessary for industry to increase its total output accordingly. The combined total production of air-separation products which are available for purchase throughout the U.S. is shown in Figure 5. Capacity is measured in tons per day, and includes both oxygen and nitrogen. In addition to the plant capacity shown, there are numerous small, independent producers of air-separation products whose output capacity is not known or included.

III. MANUFACTURE OF CRYOGENIC LIQUIDS

Several methods are available to produce cryogenic liquids such as oxygen, nitrogen, and argon. This discussion will be confined to these three cryogens because they constitute the largest portion of the cryogenic liquids in use today. Although other liquefied gases such as hydrogen, helium, neon, and krypton have boiling temperatures well within the range of crygenics, they will not be described because of their relatively small volume demand as compared to the other three.

Atmospheric air is a mixture of gases (Table 1) which under specific conditions can be separated to form individual gas elements. For practical purposes, air can be considered to be a mixture of nitrogen (78.1%) and oxygen (20.9%). The remaining 1%

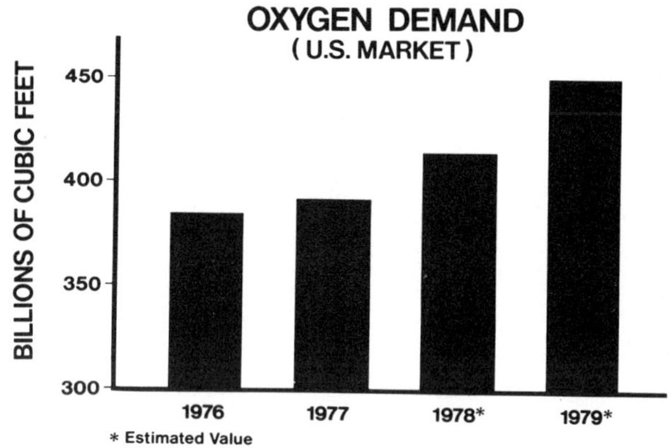

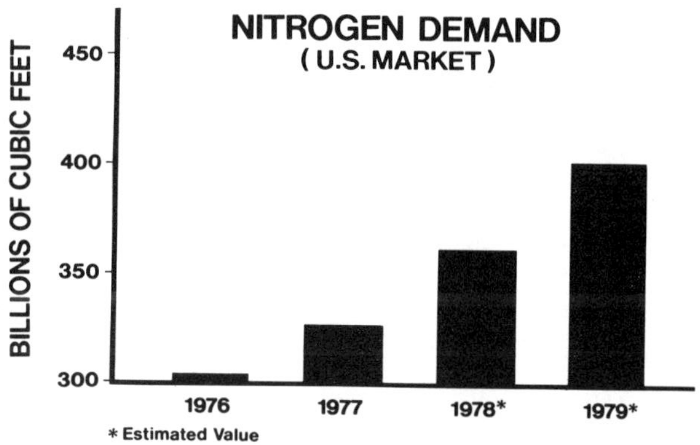

FIGURE 4. Oxygen demand — U.S. market (top). Nitrogen demand — U.S. market (bottom).

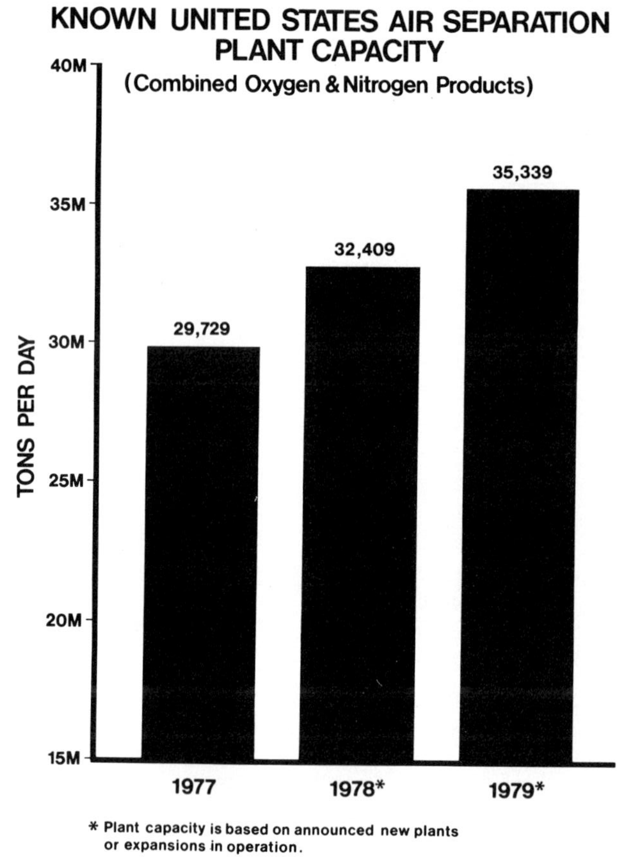

FIGURE 5. Known U.S. air-separation plant capacity (combined oxygen and nitrogen products). Plant capacity is based on announced new plants or expansions in operation.

Table 1
COMPOSITION AND PROPERTIES OF AIR

Component	Symbol	Volume (%)	Boiling point °F	Boiling point °C	Critical temperature °F	Critical temperature °C	Triple point °F	Triple point °C
Nitrogen	N_2	78.10	−320.4	−195.8	−232.8	−147.1	−346.0	−210.0
Oxygen	O_2	20.94	−297.4	−183.0	−181.1	−118.4	−361.9	−218.83
Argon	Ar	0.93	−302.58	−185.88	−188.5	−122.5	−310.81	−190.45 @ 0.68 atm
Carbon dioxide	CO_2	0.03	−109.3[a]	−78.5	87.8	31.0	−69.9	−56.6 @ 5.11 atm
Neon	Ne	18.18 ppm	−410.9	−246.07	−379.7	−228.7	−415.5	−248.6
Helium	He	5.239 ppm	−452.0	−268.9	−450.2	−267.9	−455.6	−270.9 @ 0.051 atm
Krypton	Kr	1.139 ppm	−244.5	−153.6	−82.5	−63.6	−251.3	−157.4
Hydrogen	H_2	0.5 ppm	−423.2	−252.9	−400.4	−240.2	−434.7	−259.3 @ 0.0695 atm
Xenon	Xe	0.086 ppm	−162.5	−108.06	61.3	16.3	−169.4	−111.9

[a] Sublimation point @ 1 atm.

AIR SEPARATION FLOW DIAGRAM

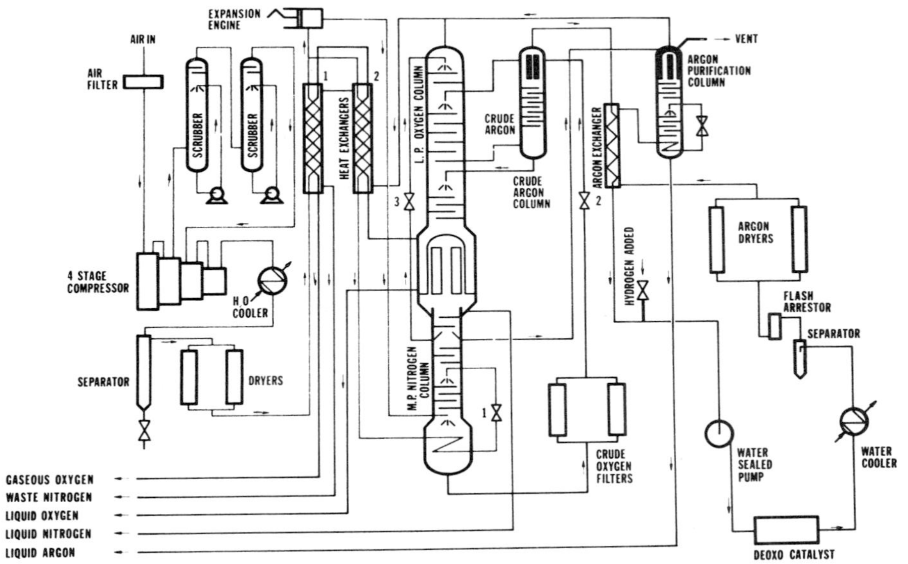

FIGURE 6. Principal components of a typical high pressure air separation plant which produces mainly liquid products and some gaseous oxygen. (Courtesy of Liquid Carbonic Corporation, Chicago.)

by volume consists of gases which are commonly referred to as the rare gases. Most air-separation plants collect the argon gas but make no attempt to capture the remaining rare gases such as neon, krypton, and xenon.

Air-separation plants can use many variations of refrigeration and process cycles. However, the process that is used is dependent upon the product mix desired, the ratio of liquid-to-gas production required, and the expected purity of the product. Plants which principally produce liquid product require greater investment, more electrical power, and higher operating pressures than plants which make only gaseous products or lower purity gases.

The separation of liquefied air using low-temperature procedures requires two basic steps which are common to all air-separation plants:

1. The continuous cooling of incoming gaseous air through a series of compressions and expansions until the air liquefies
2. The separation (or distillation) of the liquid air into the separate gases through the use of distillation column trays.

Figure 6 is a flow diagram which illustrates the principal components of a typical high-pressure air-separation plant to produce mainly liquid products and some gaseous oxygen. What follows for the remainder of this chapter is a detailed discussion of this flow diagram.

A. Basic Liquefaction Steps

1. Compression

Atmospheric air is drawn through a dust-removal filter into a four-stage air compressor. Four stages are used to minimize compression ratios and compressed air temperatures. The air is cooled between each stage in water-cooled exchangers to remove the heat of compression. Special oils are used for lubrication in the cylinders so that the air will not be contaminated with hydrocarbons.

2. Scrubbing

Between the second and third stage of the air compressor, and following the second stage aftercool, the air is passed upwards through two cylindrical vessels connected in series and packed with contact rings over which caustic soda is circulated. These towers remove all carbon dioxide from the air so that it will not collect as solid CO_2 at the low temperatures prevailing in the air-separation column and plug the system.

3. Drying

Following the compressor's fourth stage aftercooler, the air may be further cooled by a freon or ammonia refrigeration unit. The next step in the purification of the air is to remove all moisture before the air enters the insulated casing containing the distillation column and accessories ("cold box"). A moisture separator removes all free water or condensate and the air then enters one of the two driers. The driers are packed with activated alumina, which absorbs several times its weight in moisture. While one drier is in use, or "on stream", the second drier is being reactivated with heated waste nitrogen to dry it out so that it will be ready for the next cycle. Instrumentation and analytical procedures monitor the purification and drying of the "process air" to eliminate carbon dioxide and water contamination.

4. Cooling

The purified air, at 2500 psig, now enters the Number 1 heat exchanger in the cold box where it is cooled to about 34.4°C (−30°F) by the effluent cold gases from the column. The air in this exchanger passes through coiled tubes wound into a bundle and countercurrent to the cold waste nitrogen in the exchanger shell. Another coiled tube bundle contains gaseous oxygen so that it also is warmed up to near the entering air temperature by cooling the process air.

5. Expansion

After the Number 1 heat exchanger, the air stream divides. About 30% goes to the Number 2 exchanger, and 70% goes to an expansion engine. In the engine, the air is expanded from 2500 psig against a piston to an exhaust pressure of 75 psig. This results in a release of the energy built up during compression; thus a large drop in air temperature occurs. The power developed in the engine is used to drive an electric generator, and the electricity is connected back into the plant power system for recovery. The air in the engine exhaust at 75 psig will now be at about −165°C (−265°C) and is piped into the bottom of the medium pressure (M.P.) column.

B. Separation of the Liquid Air

The main column is in two sections: the lower or medium pressure (M.P.) and the upper or low-pressure (L.P.) column. This is frequently referred to as a double-rectification column because the final air separation is accomplished in two stages.

The 30% of the air which goes through the Number 2 heat exchanger at 2500 psig is cooled to about −173.3°C (−280°F) and is then expanded through the main expansion valve (1) after having been partially condensed in the coil in the bottom of the M.P. column. After this expansion with its cooling effect, the air is in liquid form as it enters the M.P. column.

In order to understand how air separation takes place, it is necessary to examine the temperatures at which the products boil at various pressures. Nitrogen has the coldest boiling temperature, which is −195.7°C (−320.4°F) at atmospheric pressure. Oxygen boils at −183°C (−297.4°F) at atmospheric pressure, and argon at −185.7°C (−302.4°F).

1. Medium-Pressure Separation System

Liquid air collecting in the lower part of the M.P. column is boiled by the coil in the bottom (through which the cold air from exchanger Number 2 is passing) and from the expansion engine exhaust stream. This boiling tends to concentrate oxygen in the bottom of the M.P. column, thus driving the nitrogen vapor up through the column trays. Each column contains 60 to 90 trays, each having many "bubble caps" which allow vapor to bubble up through the liquid which has collected on each tray. Liquid overflows a dam on each tray to flow down to the next lower tray. Thus vapor goes up the column in intimate contact with the liquid on each tray. The liquid on each tray condenses some of the oxygen and argon vapors which have higher boiling temperatures, and, in so doing, some of the liquid nitrogen liquid is vaporized by having been warmed by the oxygen and argon. As the liquid progresses down from tray to tray, it became richer in oxygen and argon. As vapors go up through the column, they become richer in nitrogen. By the time the remaining vapor has reached the top of the M.P. column, its oxygen content has been reduced to 8 to 10 ppm and is considered pure nitrogen. Since the M.P. column is controlled at 75 psig, nitrogen entering the tubes of the reboiler is condensed by the liquid oxygen which has collected in the bottom of the L.P. column and surrounding the tubes of the reboiler. This is possible because low-pressure oxygen is at −183°C (−297.4°F) and is cold enough to condense nitrogen at 75 psig, at which pressure it boils at only −175.5°C (−284°F) instead of at −195.7°C (−320.4°F). Liquid nitrogen product is collected in a dam below the reboiler tubes at the top of the M.P. column. Some of this liquid nitrogen overflows to the lower trays of the M.P. column thus helping to condense any oxygen vapor trying to ascend and in so doing becomes vaporized again and ascends. The liquid which collects in the bottom of the M.P. column is rich in the higher boiling oxygen and argon. This so-called "crude oxygen" (which is about 45% oxygen) becomes the feed to the L.P. column.

2. Crude Oxygen Filtration

The crude oxygen is first filtered in silica gel filters for removal by adsorption of trace hydrocarbons. It is then expanded through Number 2 valve and fed to the L.P. column, after first passing around the tubes of the condenser on the top of the crude argon column, which will be explained later.

3. Low-Pressure Separation System

As the crude feed enters the L.P. column, the oxygen and argon tend to flow down as liquid, while the nitrogen rises as a vapor. In order to prevent oxygen and argon from going up and out the top of the L.P. column, some liquid nitrogen from a collecting dam of the M.P. column is fed into the top of the L.P. column after being expanded through valve Number 3 to near atmospheric pressure and −195.7°C (−320.4°F). The cold liquid nitrogen ("reflux") flows down from tray to tray, condensing oxygen and argon vapor which are trying to ascend. The nitrogen is vaporized by the warmer oxygen and argon and rises to the top of the L.P. column. The nitrogen vapor is called "waste nitrogen" because it contains up to 1% oxygen. It passes through the heat exchangers Number 2 and Number 1 where it is used to cool the incoming process air after which it is wasted to the atmosphere.

4. Crude Argon Separation

As the liquid in the L.P. column flows down through the bubble cap trays, it gradually warms and becomes richer in oxygen and argon as the nitrogen content is vaporized and flows upwards. At a point between the "crude oxygen" feed and the reboiler section at the bottom of the L.P. column, a point is reached where the argon concentration is maximized and nitrogen has boiled off. At this point, a side stream of vapor oxygen and argon is directed to the crude argon "side arm" column. In this column, argon having a boiling temperature colder than oxygen goes up through bubble cap trays as a vapor while the oxygen content is washed down and becomes condensed in the descending liquid oxygen. This oxygen is fed back into the main L.P. column. Any oxygen trying to go out the top of the crude argon column is condensed in the tubes of the condenser which are surrounded by colder crude oxygen liquid.

The liquid which reaches the bottom of the L.P. column is the oxygen which is drawn off as product. Vapor oxygen may also be withdrawn from a point just above the liquid oxygen level.

5. Argon Purification System

The argon vapor from the top of the crude argon column still contains 1% or 2% of oxygen which must be removed. This vapor is warmed in an exchanger in which warm purified argon from the oxygen removal (Deoxo®) equipment passes countercurrently. The warm crude argon is continuously analyzed for oxygen content, and the analyzer automatically operates a controller which adds the correct amount of hydrogen which combines with the oxygen and converts it to water in the Deoxo catalyst. The oxygen-free argon is then dried in an alumina-filled drier and cooled in the crude argon heat exchanger. From here it goes to the final argon purification or redistillation column where pure argon collects as liquid at the bottom, and traces of nitrogen and hydrogen go out the top past the condenser tubes which are cooled by liquid nitrogen.

6. Quality Control

In order to control this process, temperatures and pressure must be maintained, and liquid levels in the condensers and reboiler must be correct. Automatic instrumentation does this in connection with analyzers for each step of the process. In this way quality control of all three products is assured. Recording analyzers continuously monitor the purity of products.

Since it takes up to 16 hr to reach the cold operating temperatures after a warm start-up, obviously air separation should be a continuous process. It takes over 24 hr before complete equilibrium is reached for maximum argon production.

IV. OTHER AIR-SEPARATION PLANT SYSTEMS

Several variations of the described plant system in Sections A and B are used. One type is similar to the liquid plant indicated in Figure 6 except that no appreciable oxygen or nitrogen liquid can be withdrawn and the oxygen production is in gaseous form. These plants operate at compressed air pressures of only 300 to 500 psig because less refrigeration is required. They require a gas holder, an oxygen compressor, a high pressure water separator, and a drier in order to fill cylinders.

Another variation of the illustrated flow diagram is a plant in which an expansion engine is not used. These plants operate at 1500 to 2000 psig compressed air pressures because the refrigeration obtained by expansion through the Number 1 air separation valve is less efficient than expansion through an engine. These plants are generally small ones producing less than 3500 ft^3 or 100 m/hr. They also require a gas holder, oxygen compressor, and drier to fill cylinders. A third variation is a so-called "liquid pumper". A high-pressure liquid oxygen pump and/or a liquid nitrogen pump are located where the drawing shows "liquid oxygen out" and "liquid nitrogen out". These are hydraulic plunger pumps which pressurize the liquid products to cylinder or tube trailer filling pressure. The high-pressure liquids flow through separate coils in the heat exchangers so that the refrigeration of the liquids is recovered by cooling the incoming air, and the liquids are vaporized to high-pressure gas products directly for cylinder and tube filling. These plants operate at air pressures between 600 and 1200 psig, and do not require gas holders, oxygen and nitrogen compressors, or driers. The newer plants for production of products in gaseous form are generally of this type.

REFERENCES

1. **McClintock, M.,** *Cryogenics,* Reinhold Publishing, New York, 1964.
2. **Scott, R. B.,** *Cryogenic Engineering,* D Van Nostrand, Princeton, N.J., 1959.
3. **Sittig, M.,** *Nitrogen in Industry,* D Van Nostrand, Princeton, N.J., 1965.
4. **Vance, R. W. and Duke, W. M.,** *Applied Cryogenic Engineering,* John Wiley & Sons, New York, N.Y., 1962.
5. **Sittig, M. and Kidd, S.,** *Cryogenics Research and Applications,* D Van Nostrand, Princeton, N.J., 1963.
6. **Braker, W. and Mossman, A.,** *Matheson Gas Data Book,* 5th ed., Matheson Gas Products, E. Rutherford, N.J., 1971.
7. Liquid Carbonic Corporation, Manufacturing of Oxygen, Nitrogen, Argon, Tech. Sales Bull. No. IG-2017, Liquid Carbonic Corp. Chicago, 1973.
8. Liquid Carbonic Corporation, Physical and Chemical Gas Data Chart, Form No. 6577, Liquid Carbonic Corp. Chicago, 1976.

Chapter 5

FRACTURE TOUGHNESS OF MATERIALS AT CRYOGENIC TEMPERATURES

J. A. Koutsky

TABLE OF CONTENTS

I. INTRODUCTION

Fracture toughness (the amount of energy needed to produce fracture in a material) is an important consideration in determining substances which can be cryopulverized efficiently. Impact-testing is one of the most common methods to measure fracture toughness and has the advantage of more accurately reproducing the impact stresses which occur during pulverization in a mill. Figure 1 shows qualitatively how fracture toughness of typical materials changes as a function of temperature. The three basic types of material behavior are

1. Those easily fractured at −196°C (−319°F) [(type A where $T_{bA} > -196°C$) (−319°F)]
2. Materials which remain ductile below −196°C, (−319°F) [(type B where $T_{bB} <$ −196°C) (−319°F)]
3. Materials which have no sharp embrittlement temperature (type C)

Many materials show a distinct brittle temperature which depends on molecular properties of the material. At this temperature, the yield stress exceeds the fracture stress of the material. The brittle temperature measurement is also subject to how fracture toughness is measured. Commonly, rapid deformation tests (high strain rates or impact tests) give increased brittle temperatures of the materials when compared to slow tests.

Cryogenic processing of materials in the "C" category is difficult to predict since there is no sharp brittle temperature. The maximum level of fracture toughness which could be handled easily by a hammermill would be about 13.5 J (10 ft/lb). If the material exhibits a fracture toughness below this value, it probably can be efficiently processed.

When impact-tested, most materials show different brittle temperatures depending on the type of test. Notched specimens give higher temperatures of the brittle-ductile transition when compared to ordinary specimens. Any specimen geometry which induces large internal stress gradients, such as notches, sharp edges, etc. will generally give higher brittle-ductile transitions. For instance, a common semikilled mild steel will show a brittle transition temperature of 5°C (41°F) by a Charpy "V" impact test, while a Charpy keyhole impact test will give a brittle transition temperature of −30°C (−21°F) for the same material at approximately the same rate of impact. The Charpy "V" test inherently produces greater stress gradients in the material. In hammermill operations for producing granulated materials, the higher quoted values of temperature for the brittle transition should be of greatest interest since microcracks induced in the specimen by impact provide large stress gradients.

II. PROPERTY VARIABLES AFFECTING FRACTURE TOUGHNESS

A. Polymers and Plastics

In order to understand why certain polymeric materials will fracture easily while others will not (even though the materials appear to be similar in many respects at room temperature), qualitative influence of molecular properties and additives in common polymeric materials on fracture toughness will be discussed. General rules which determine fracture toughness are shown in Table 1. Table 2 shows several common polymers which have been tested for fracture at liquid nitrogen temperatures.

The first consideration is the glass transition temperature (brittle temperature) which is determined for linear, branched, or lightly crosslinked polymeric materials and is

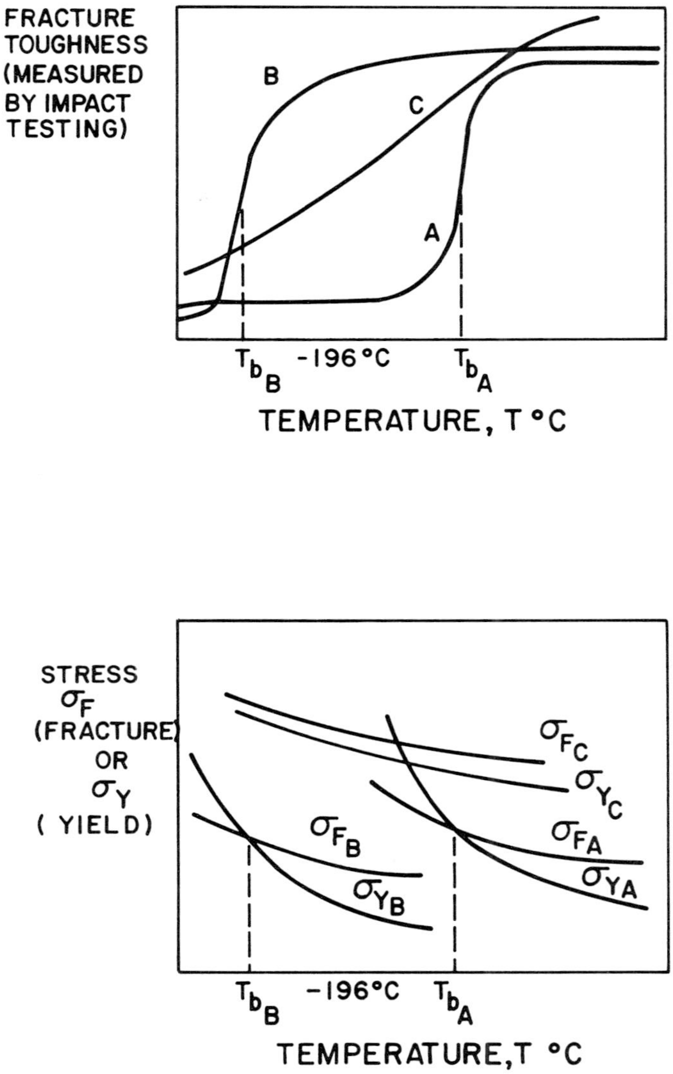

FIGURE 1. Qualitative effects of temperature on fracture toughness and yield and fracture stresses for three different types of materials.

quoted widely in the literature for many of these materials. This property is determined exclusively by the amorphous content of the polymer. If the polymer can crystallize, then the fracture toughness is not only governed by the glass transition temperature but also by the percent crystallinity and how the crystals are distributed in the polymer.

Crystallinity can have an enormous influence on the fracture properties of a polymer. For example, polyethylene terephthalate is a material that can be found in a variety of forms. Its fracture toughness is considerably increased if the polymer is either in a fiber form or in a heat-treated film form; thus it cannot be fractured easily even at liquid nitrogen temperatures. This is observed even though the glass transition is quoted at 69°C (156°F) in the literature.

Polarity is also an important consideration. Normally the higher the polarity of the material, the more difficult it is to fracture. A good example of this are the polyamides (nylons) which have high crystallinity and are highly polar.

Table 1
INFLUENCE OF MOLECULAR PROPERTIES AND ADDITIVES ON FRACTURE TOUGHNESS OF POLYMER MATERIALS

Molecular properties	Influence on fracture toughness (ft)
Glass transition↑	↓
% Crystallinity↑	↑
Polarity↑	↑
Molecular weight↑	↑
Cross-link density↑	↓
Additives	
% Plasticizers↑	↑
Fillers	
% Inert particles↑	↓
% Rubber↑	↑
% Fibers↑	↑

For linear and branched polymeric materials, the influence of molecular weight on fracture toughness can be considerable. The incorporation of small weight percents ($\sim$1%) of ultra-high molecular weight species (>10^6) can increase fracture toughness as much as three to ten times the normal values.

As the crosslink density of a material increases, the fracture toughness will begin to also increase. However, normally at even low degrees of crosslinking density, the fracture toughness will begin to be considerably reduced if additional crosslinks are added. This can be accomplished through irradiation by high-energy gamma rays, high-intensity ultraviolet rays, or by partial chemical degradation.

Additives can have a profound influence on the fracture toughness of a polymeric material. As indicated in Table 1, the use of plasticizers (usually incorporated in vinyl resins) increases fracture toughness by lowering the glass transition temperature. Generally, inert fillers decrease fracture toughness while fibers will increase fracture toughness. For inert fillers it has commonly been observed that a linear relationship exists between fracture toughness and volume percent filler.

B. Copolymers — Terpolymers

The fracture toughness of copolymers and terpolymers is strongly dependent on the types of monomers used and their relative percentages in the material. If one of the monomers is a "stiff" molecule, the resulting brittle transition will be increased even though only small percents of the stiff molecule are utilized. The sequential distribution of the monomers is also important in determining fracture toughness. Thus copolymers can have random, alternating, block or graft sequences all having the identical percent composition of monomers, but each copolymer can have entirely different values of fracture toughness.

C. Metals and Metal Alloys

As with polymers, metals exhibit decreases in fracture toughness with increasing stress or strain rate. Additionally, the brittle temperature shifts upward, as is shown in Figure 2. The brittle temperature can be defined either arbitrarily by the impact behavior using a specific test or more fundamentally as the temperature at which the yield stress is equivalent to the fracture stress (i.e., when no yielding of the metal is

Table 2
POLYMERS STUDIED FOR FRACTURE IN LIQUID NITROGEN

A. Polymers Which Fracture at Liquid Nitrogen Temperatures

Polymer	Glass transition temperature (°C)	Remarks
Polyethylene (low density)	−25	High crystallinity, low polarity
Polypropylene	0—20	High crystallinity, low polarity
Polyvinyl chloride	105	Low crystallinity, polar
(plasticized)	<105	Low crystallinity, polar
Polystyrene	100	Low crystallinity, nonpolar
Styrene-butadiene	−70, −100	Low crystallinity, nonpolar
All highly cross-linked thermosets (phenolics, epoxies, melamines, ureas, alkyds)	Degrade before softening	High cross-link density, non-crystalline, easily fractured

B. Polymers Which Do Not Fracture at Liquid Nitrogen Temperatures

Polymer	Glass transition temperature (°C)	Remarks
Polyethylene fiber	−25	High crystallinity, high orientation
Polyethylene terephthalate	69	
Film	69	Intermediate crystallinity, polar
Fiber	69	High crystallinity, polar, oriented
Polytetrafluoroethylene	126	High crystallinity
Polyhexamethylene adipamide	50	High crystallinity, polar

Table 2 (continued)
POLYMERS STUDIED FOR FRACTURE IN LIQUID NITROGEN

C. Fracture Characteristics of Various Commercial Plastic Samples Cooled by Liquid Nitrogen

Polymer Sample	Remarks on ability to fracture
Homogeneous—films, sheet, and tubing	
Low density polyethylene film (2 mm thick)	Difficult to fracture
Low density polyethylene tubing (½ in. diam., ⅛ in. thick wall)	Easy to fracture
Polypropylene film (2 mm thick)	Fractures easily if kept close to liquid nitrogen temperatures
Polyvinyl chloride-plasticized tubing (Tygon®)	Easy to fracture
ABS (Cycolac®) sheet (¼ in. thick)	Easy to fracture
Polytetrafluoroethylene (Teflon®) film (3 mm thick)	Will not fracture
"FEP" fluorocarbon film (3 mm thick)	Will not fracture
Surlyn® (Polyionomer) (10,4,2, mm thick)	Easy to fracture
"Nylon 6" film ½ mm thick)	Will not fracture
"Nylon 6" film (2 mm thick)	Will not fracture
Polyethylene terephthalate (Mylar®) (2 mm thick)	Will not fracture
Composite Films	
Biaxially oriented polypropylene (2 mm) — Saran® (½ mm), EVA copolymer (2.5 mm)	Fractures easily if kept close to liquid nitrogen temperatures
Polyethylene (2.5 mm) — Saran (½ mm), "Nylon 6" (1 mm)	Fractures easily if kept close to liquid nitrogen temperatures
Polyethylene (2 mm) — Saran (½ mm), semi-rigid vinyl (10 mm)	Easy to fracture (film splits)
EVA copolymer (1.25 mm) — Saran (½ mm), semi-rigid vinyl (10 mm)	Easy to fracture (film splits)
Foams	
Cellulose acetate foam (1 in. thick)	Easy to fracture
Polyurethane foam (5 in. thick)	Easy to fracture
Fibers	
"Nylon 66"	Will not fracture
Nylon tire cord	Will not fracture
Dacron®	Will not fracture

observed). For the purpose of cryopulverization, high stress or strain rates should be employed to approximate the process in common impact mills.

For pure metals the crystallographic structures of the unit cells are important in determining their brittle points. Table 3 indicates the various crystallographic structures of some metallic elements. It has been observed that FCC metals show no significant brittle point and are widely used for low-temperature structural applications. This is due to less inhibited slip and dislocation motion. BCC metals, however, exhibit characteristic brittle temperatures below which a large decrease in fracture toughness

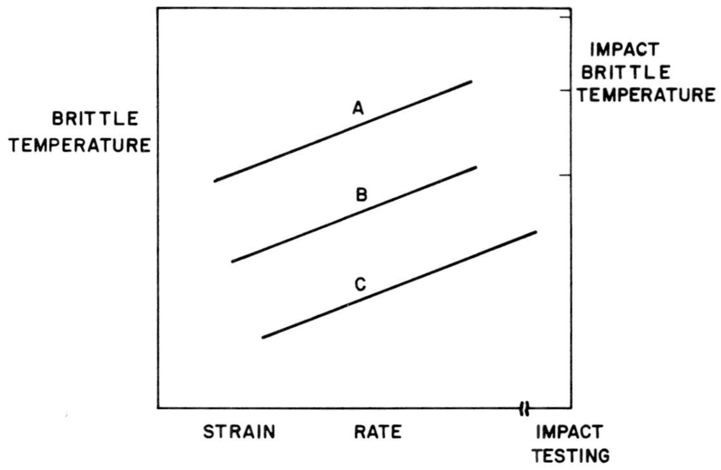

FIGURE 2. Effect of deformation rate on observed brittle temperature where transition to brittle behavior is observed.

Table 3
CRYSTALLOGRAPHIC STRUCTURE OF METALS

Face-centered cubic (FCC)	Bodied-centered cubic (BCC)	Hexagonal close-packed (HCP)
Aluminum (Al)	Chromium (Cr)	Beryllium (Be)
Copper (Cu)	α-Iron (Fe)	Cadmium (Cd)
Gold (Au)	Molybdenum (Mo)	Magnesium (Mg)
γ-Iron (Fe)	Tantalum (Ta)	Titanium (Ti)
Lead (Pb)	Tungsten (W)	Zinc (Zn)
Nickel (Ni)	Vanadium (V)	
Platinum (Pt)		

is observed. Figure 3 shows typical qualitative behavior of FCC and BCC metals. HCP metals can be either brittle or ductile at low temperatures. There is no general rule for these materials.

Impurities and alloying elements can have considerable influence on the fracture toughness properties and brittle temperature of a metal. Generally, impurities which form hard intergranular precipitates such as carbide, hydrides, nitrides, sulfides, and oxides reduce fracture toughness. On the other hand, alloying elements which can remove or fix oxygen, sulfur, or nitrogen, for example, prevent precipitate formation and considerably increase fracture toughness. For iron-based steels, fracture toughness is increased by the addition of such elements as aluminum, titanium, and zirconium.

Grain size and texture have a large influence on fracture toughness. Fine-grained metals are usually much tougher at low temperatures than coarse-grained metals. Grain size can be controlled by thermal treatments and the addition of alloying elements. Texture or the continuity of the phases is primarily influenced by thermal and mechanical treatments. For instance, in pearlitic steels, if the brittle ferritic phase is continuous, a steel of low-fracture toughness is observed. Table 4 summarizes typical fracture data for several common metallic materials.

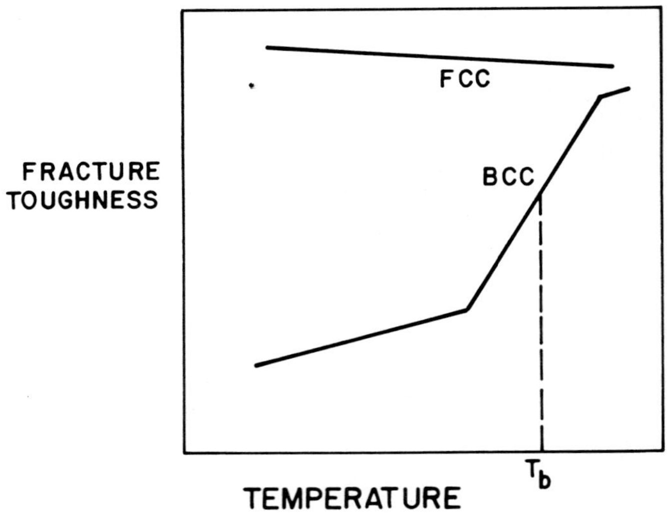

FIGURE 3. Typical qualitative behavior of FCC and BCC metals.

Table 4
FRACTURE DATA FOR COMMON METALLIC MATERIALS

Ductile at liquid nitrogen	Brittle at liquid nitrogen
Cast aluminum	Steel SAE 1020
Copper wire	Steel CR 1045
Cast bronze 80-10-10	Stainless steel 410
Brass ASTM B21	Cast iron (ductile)
Steel SAE 1045	Cast iron (grey)
Steel HR 1045	Cast iron (malleable)
Steel SAE 6165	Lead base Babbitt
Stainless steel 303	Tin base Babbitt
Stainless steel 304	
Stainless steel 316	
Stainless steel 416	

D. Thermal Expansion Effects

Rapid quenching of a composite material often produces cracks due to differential thermal coefficients of expansion of the phases comprising the material. These cracks readily form at the interfaces between the phases and can lead to subsequent interfacial failure. This process can aid in separating constituents such as in the case of laminated materials (i.e., polymer-polymer and polymer-metal laminates or polymer-coated metals such as wiring, cabling, and painted surfaces). Generally, one of the phases must remain ductile or have a high fracture toughness at the lower temperatures in order for the process to work efficiently in the cryopulverizer.

III. CRYOPULVERIZATION OF MATERIALS — CONCLUSION

The cryopulverization of materials produces large surface areas of the material. Normally the freshly produced surfaces are inactive at liquid nitrogen temperatures and in the presence of nitrogen. However, subsequent handling of the material can be crucial for successful processing since the temperature of the pulverized material increases, and chemical and physical reactions can occur.

In the case of metals, reaction with oxygen could be violent or even explosive. These metal powders should be carefully handled, and contact with air should be carefully controlled.

For certain lightly crosslinked, rubbery polymeric materials, agglomerization can readily occur due to adhesion and molecular flow between the particles. This can be prevented if the particles are precoated with a talc or inert surfactants before increasing the temperature above the brittle point.

Chapter 6

EQUIPMENT FOR CRYOGENIC PROCESSING

R. Kisielewski

TABLE OF CONTENTS

I. INTRODUCTION

In recent years there has been a greater awareness for the conservation of our natural resources. As a consequence, the recycling of discarded consumer goods and other solid wastes is becoming a necessity to help alleviate the shortages of virgin feed stocks. Although the aims of recycling are admirable, the cost of reclaimed materials in many instances is higher than the price of their virgin counterparts. New processing techniques are being developed to reduce the cost of reclaimed materials to make them competitive in the markets.

Within the last decade cryogenic technology has been applied to the reclaiming of materials which were previously difficult to process. Cryogenic processing is based on the principle that many materials exhibit brittle characteristics when their temperatures are reduced to low levels. At these low temperatures the materials behave like glass and consequently can be fractured readily. The highest temperature at which a material exhibits this behavior is called its cold cracking temperature or brittle point. Rubber which is ductile at room temperatures will fracture like glass when its temperature is reduced to about −62°C (−80°F). All materials do not behave in this manner. Aluminum and copper retain their ductile characteristics down to the temperature of liquid helium −267°C (−450°F).

Cryogenic processing relies on the preceding principles to liberate the various fractions from a scrapped item. Components such as electrical transformers can be frozen to embrittle the iron core laminations, then shattering them and separating the iron fragments from the ductile copper windings. The reclaimed material is of a high quality yielding bright shiny copper and cleanly broken fragments of iron. Incineration is an accepted method of reclaiming copper wire from electrical conductors and components such as solenoids and relays; however, the copper becomes oxidized and is covered with a carbonaceous residue which degrades the copper to a number 2 or lower grade. Cryogenic processing has the advantage of yielding a high quality product. A few of the items processed with the cryogenic technique are: automotive-type tires, plastic wires, starters, generators, electric motors, plastic-metal laminates, electrical conductors, and plastic bleedings.

The main criterion for cryogenic processing is that one or more of the components in an item to be processed must have an attainable cold cracking temperature. The materials can then be fragmented, separated, and reclaimed.

II. FREEZING SYSTEMS

In cryogenic processing the scrapped items must have their temperatures reduced to their respective cold cracking temperatures just prior to being fed into a fragmentizer. This task is accomplished by subjecting the items to a cryogenic environment for a period of time sufficient to attain this temperature. This environment can be in the form of a liquid, gaseous, or liquid-gaseous media. The simplest form of freezing is by immersion of an item into a liquid cryogen such as liquid nitrogen (−195°C [−320°F]).

A. Immersion Batch Freezing

Styrofoam, polyurethane, or insulated stainless steel or aluminum tanks can be used to contain the cryogen for the batch immersion freezing method. Items can be lowered into the liquid for a time period sufficient to reduce the temperature of the item to its cold cracking temperature, then removed and fragmented. Items can be left in the cryogenic bath until all boiling ceases, at which time the temperature of the object will be that of the liquid cryogen. Extreme caution must be exercised when using this

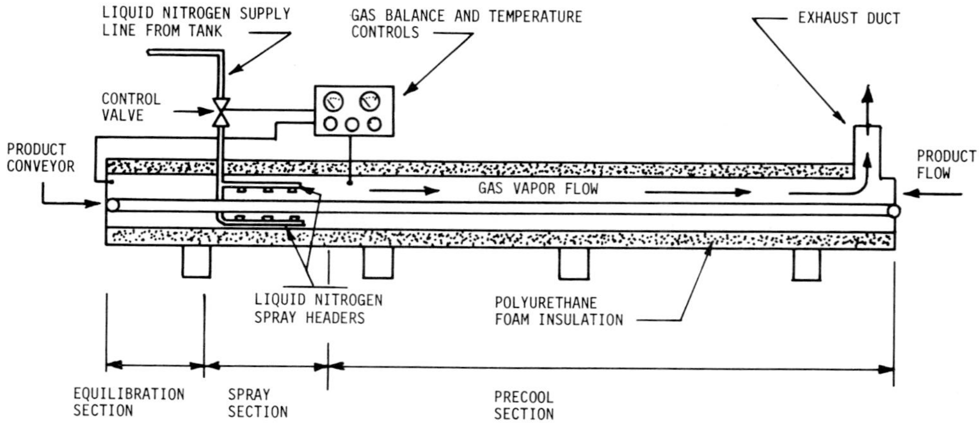

FIGURE 1. Cross-sectional view of a typical liquid nitorgen-freezing tunnel.

method since violent boiling of the liquid occurs when the material is first lowered into the cryogen. Cryogenic liquids which come in contact with the human skin cause severe freeze burns; therefore, proper safety procedures must be strictly adhered to.

Batch immersion processing has limited applications since processing throughout is extremely low. In addition, the processing equipment is subjected to intermittent loading conditions which reduce overall processing efficiency.

B. Liquid Cryogen Freezing Systems

The most efficient utilization of a cryogen is obtained by using a freezing tunnel where the enthalpy of both the liquid and vapor phases is utilized for freezing. As a product moves continuously through the tunnel, it is initially precooled by the cryogenic vapor; then near the discharge end of the tunnel the product is sprayed with liquid cryogen for the final cooldown. The sprayed liquid vaporizes as it extracts heat from the product. This vapor which is at the same temperature as the liquid cryogen is directed down the tunnel toward the feed end. Heat energy is removed from the product as this vapor progresses over the incoming product in a counterflow pattern. Upon leaving the freezing section, the product is directed immediately into a hammermill for fragmentization.

A typical liquid nitrogen freezing tunnel system is illustrated in Figure 1. The product which enters the tunnel from the right must be uniformly distributed in a thin layer (burden height) over the moving conveyor. Voids on the conveyor belt lower the freezing efficiency of the system. Greater efficiency would be obtained, for example, by processing 2 in.^{2} tire chips which can be distributed over the belt in a solid uniform layer as opposed to whole tires where the belt would contain approximately 40% of unused space. Voids on the conveyor belt increase the per-pound freezing cost since all of the usable enthalpy of the vapor phase is not utilized for freezing.

As the product progresses deeper into the tunnel, it is constantly being subjected to a colder environment which continues to cool the product. The vapor phase of liquid nitrogen removes 0.25 Btu of heat energy from the product for every 0.56°C (1°F) of temperature change. The temperature of the product is reduced to its cold cracking point in the spray section located near the discharge end of the tunnel. Generally, liquid nitrogen is sprayed over the product from an overhead spray bar; however, an additional spray bar can be incorporated in the system to also spray the product from the bottom. The change of stage of the liquid nitrogen from the liquid to the gaseous phase removes 86 Btu of heat energy from the product for each pound of cryogen

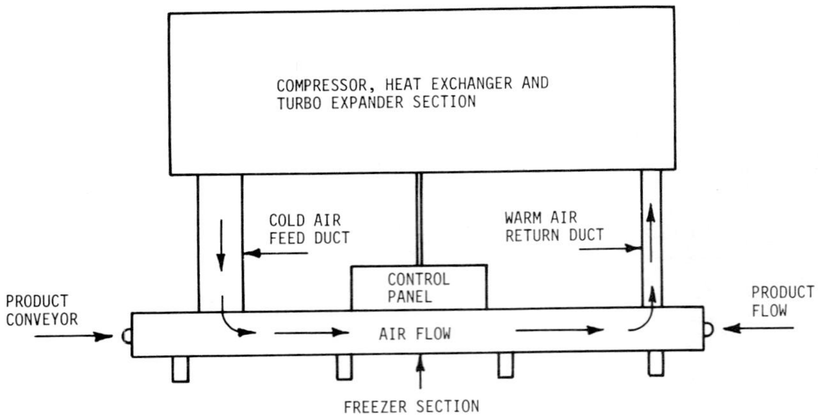

FIGURE 2. Schematic of a cold air freezing system.

consumed. The vapor phase which is at −195°C (−320°F) is then directed upstream to precool the incoming product.

The area downstream from the spray section is used to allow the temperature in the product to equilibrate. The larger the mass of the product the longer the required equilibration period. A typical equilibration time for frozen scrap tire chips is approximately 10 sec. Gas balance and temperature controls are utilized in these systems to regulate the flow of the liquid nitrogen in order to prevent the exhausting of excessively cold vapors. Discharging of excessively cold vapors into the atmosphere results in increased freezing costs.

The temperature of the product entering the freezing tunnel has a large influence on the economics of the freezing operation. The lower the temperature of the incoming product, the lower the cryogen requirement. Cryogenic processing costs could be reduced by using a seasonal approach to the various stages of scrap reclaiming. Scrap materials could be preprocessed during the warm months and cryogenic processing postponed until the cold months. For example, in the cryogenic processing of scrap tire chips, the liquid nitrogen consumption would be reduced in half if processing were carried out during the winter months when the incoming produce is at −23°C (−10°F) as opposed to summer operation when the production is at 27°C (80°F).

C. Cold Air Freezing Systems

In areas where drayage charges for transporting liquid nitrogen would make cryogenic processing economically unfeasible, a cold air freezing system could be utilized. These units produce refrigerated air at a temperature of −129°C (−200°F) through a turbo expansion system. The −129°C air is circulated through a freezing tunnel which is similar in construction to the liquid nitrogen systems (Figure 2). These freezing units are closed loop systems (the air which is exhausted from the freezing tunnel is returned to the first stage compressor of the refrigeration system). A small amount of makeup air must be added to the system to compensate for the loss of air in the freezing tunnel.

The initial investment for cold air systems is high — over one million dollars for a unit of sufficient capacity to process 13 metric tons (18 tons) of tire chips per hour. These systems, however, appear to offer attractive economic advantages over the liquid cryogen systems where high tonnage processing is to be carried out.

Presently there are no commercially manufactured freezing systems made specifically for the scrap recycling industry. Food freezing tunnels can be adapted to the needs of the scrap processing industry although generally these tunnels have limited burden height capability. Food processing freezers are fabricated of stainless steel for both the interior and exterior surfaces because of the sanitation requirements in the food processing industry. The cost of freezing tunnels can be reduced by using stainless steel for the interior surfaces which are subjected to the cryogenic environment and carbon steels for all other surfaces.

Floor space limitations can preclude the use of a freezing tunnel since its length can range from 8.5 m to 26 m (30 to 90 ft). Spiral belt conveyor freezing systems can be used to reduce floor space requirements. These systems utilize square or rectangular floor-plan rooms which contain one or more spiral conveyors for transporting the product through the cryogenic environment. A conveyor in one of these systems spirals upwards along a verticle axis similar to a spiral staircase. The floor space requirements of a 900 kg (2000 lb)/hr freezing system can be reduced from a space approximately 1.4 m × 1.4 m × 14 m (5 ft × 5 ft × 50 ft) for a tunnel system to a space of about 2.3 m × 2.3 m × 2.3 m (8 ft × 8 ft × 8 ft).

The product enters these systems at floor level and is transported towards the ceiling where it exits at ceiling level in a single-cell unit or continues to a second spiral conveyor in a two-cell unit. The product then spirals downward to floor level where it exists from the freezer. The cryogenic environment in these systems can be provided by either liquid nitrogen, liquid carbon dioxide, or cold air.

III. SIZE REDUCTION OF MATERIALS

Following the freezing operation, a material must be fragmentized by some type of equipment. There are four mechanical methods for reducing the size of materials: attrition, shearing, compression, and impaction.

A. Attrition

Attrition is the physical separation of one material by another material which is harder than the first. The sharpening of a tool bit on a grinding wheel is an example of attrition by abrasion (Figure 3). The hard particles in the grinding wheel remove small steel particles from the tool. This form of processing produces a powder residue in one stage of operation; however, the process is very slow.

Another form of attrition size reduction is in the use of skiving knives to cut away small pieces of materials such as rubber or thermoplastics. The product from this process is a mixture of material ranging from dust to thin narrow slivers of the material being processed. Tire recapping plants utilize this method to prepare tire casing for recapping. The skiving knives pass over the tread face of the rotating tire, cutting away the tread face until the desired profile is attained (Figure 4). The attrition methods of size reduction have very limited applications in cryogenic processing.

B. Shearing

Size reduction by the shearing method is accomplished by passing material between two adjacent blades as they move past each other. The three basic types of shears are alligator, guillotine, and rotary (Figure 5). The first two are cyclic in nature and have limited throughputs. Rotary type shearing machinery is able to process on a continuous basis and therefore has high throughputs.

Alligator and guillotine shears are used routinely in the scrap metal recycling industry to reduce the size of scrapped items such as "I" beams, channels, pipe, and wire. Although alligator and guillotine shears are not used directly in cryogenic processing, they are, however, used to preprocess materials for cryogenic fragmentizing.

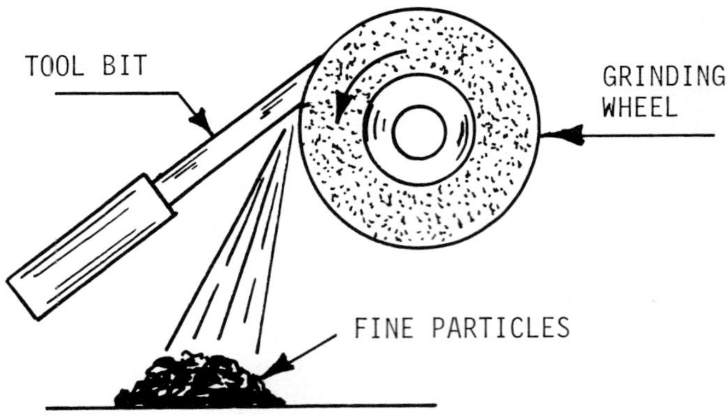

FIGURE 3. Size reduction by attrition (abrasion mode).

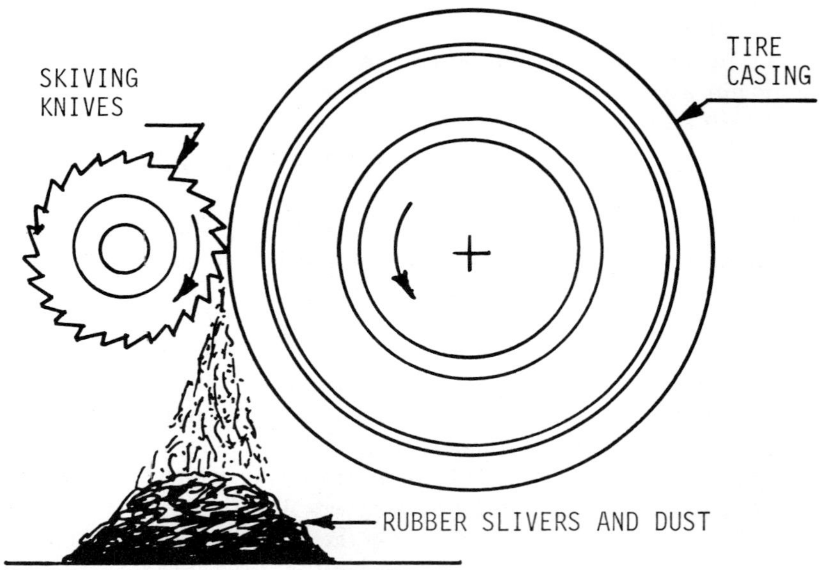

FIGURE 4. Size reduction by skiving.

In recent years the rotary shearing principle has been applied to machinery designed for the processing of materials such as scrap tires, metal skeletons, electrical conductors, and solid waste. Machinery of this type is able to reduce polymeric materials to particles of about 6.4 mm (¼ in.) in size. Although this equipment is able to produce a small product size, it is not able to liberate the various fractions in the product. Scrap tires, for example, can be reduced to 6.4 mm (¼ in.) size; however, the rubber cannot be separated from the cord or wire fractions. Size reduction to a product of 5 cm (2 in.) or smaller requires a multistage shearing operation. Generally, the items being processed are reduced to hand-size pieces in the first stage and reduced to the final size in subsequent stages.

Although rotary shear machinery is usually not directly applicable to cryogenic size reduction, it is applicable to preprocessing of materials for cryogenics. Scrap rubber or plastic products can be reduced to hand-size or smaller pieces at relatively high throughputs.

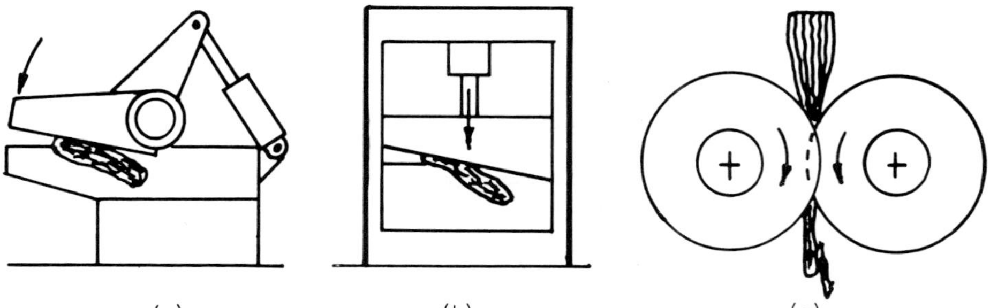

FIGURE 5. Types of shears: (a) alligator, (b) guillotine, (c) rotary.

Various equipment is available to shear scrap tires into pieces ranging in size from 6.4 mm (¼ in.) cubes to approximately 30 cm (12 in.) segments. One mobile rotary-shear tire processing plant specifically designed to preprocess scrap tires for cryogenic processing can shear tires into approximately 2 in.^{2} chips at the rate of 16.3 metric tons (18 tons)/hr. This plant is powered by a diesel engine driving hydraulic pumps which power the shearing, conveying, and screening functions. A tandem-axle tractor pulls the plant between processing sites.

The 5 cm (2 in.) chips produced by this plant can be uniformly distributed over the conveyor belt in a cryogenic freezing system, thus increasing the efficiency of the freezing cycle.

C. Compression

In the simplest form of the compression method for size reduction, brittle objects are fragmentized by applying a compressive force to the object as it rests on an anvil (Figure 6). This method of fragmentation has been applied to cryogenic processing. Embrittled materials are placed in a chamber, then a platen moves down into the chamber to compress and fragmentize the materials. Being essentially a batch processing system, this method lacks the ability to process at high throughputs. Additionally, it is difficult to regulate the particle size of the finished product.

D. Impaction

After materials which have been frozen to their cold cracking point are removed from the cryogenic environment, their temperature begins to rise. It is, therefore, essential in cryogenic processing to fragment embrittled items as quickly as possible to be assured that the processing is completed at no higher than the cold cracking temperature of the item. Brittle objects can be fragmented by subjecting them to an impact force (Figure 7). If a rock or a pane of glass is struck with a hammer, it will shatter into many pieces. Hammermills are machines which operate on this principle and have been used for decades to pulverize limestone, rocks, shale, and coal. Small hammermills are used for microgrinding materials such as nutmeg, pepper, nuts, pigments, and chemicals into a powder consistency.

In recent years large-capacity hammermills such as those used to pulverize rocks, coal, grains, and cereals have been adapted for use in cryogenic recycling of scrap materials. It is possible with this equipment to process items such as scrap tires, electrical transformers, automotive starters and generators, and plastic bleedings to reclaim the various materials of which they are composed. Hammermills are available with throughputs ranging from about 45 kg (100 lb) to 1600 metric tons (1800 tons)/hr and with power requirements ranging from 1.5 (2) to 3000 kw (4000 hp). As a

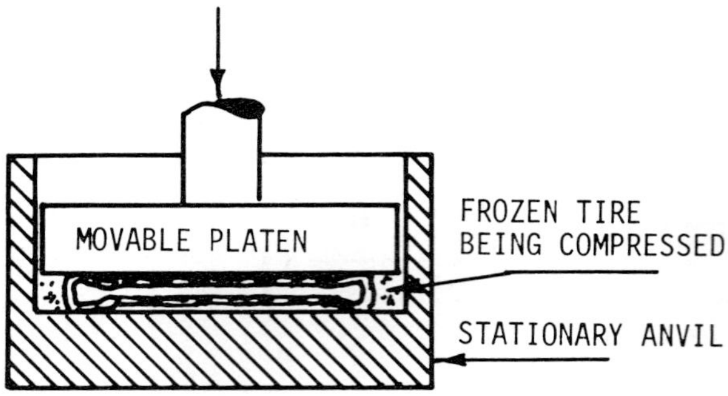

FIGURE 6. Size reduction by compressing brittle object.

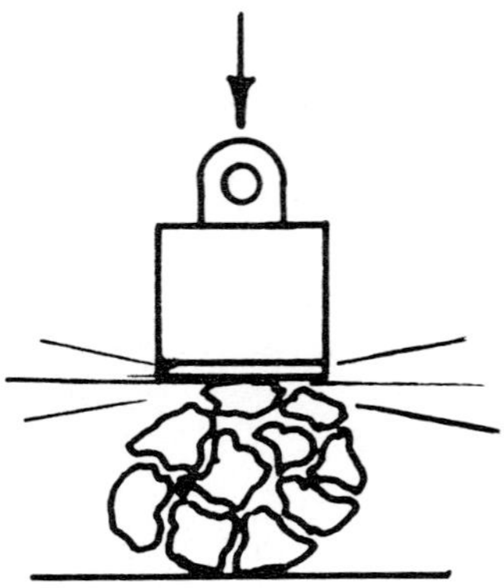

FIGURE 7. Fragmenting a brittle object by impact loading.

general rule, power requirements for cryogenic processing are lower than those for ambient temperature processing.

IV. HAMMERMILL CONSTRUCTION

A common element in all hammermills is a balanced rotor assembly which rotates at speeds ranging from 500 to 3600r/min during operation. Rotors in microgrinding hammermills may turn at speeds as high as 10,000 r/min. Figure 8 illustrates three types of rotor assemblies for use with nonrigid hammers. The backbone of the assembly is the rotor shaft onto which are mounted either rotor disks (Figure 8a), rotor plates (Figure 8b), or rotor spiders (Figure 8c). Three or more rows of hammers are pinned to the disks. Any of these rotor types can be used in hammermills for cryogenic processing. The circular disk rotor is recommended for processing large items such as electrical transformers which contain a large ductile fraction. If items such as these are not thoroughly embrittled when they enter the mill, they can cause damage to the equipment in square plate or spider equipped rotors (Figure 9). The ductile items will jam between the rotor and breaker plates inducing high stresses in the shaft, bearings, and housing. This would result in bent shafts, broken bearings, and bent frames. Rotors with circular disks are less susceptible to damage in such circumstances because the items tend to slide along the surface of the rotor disks.

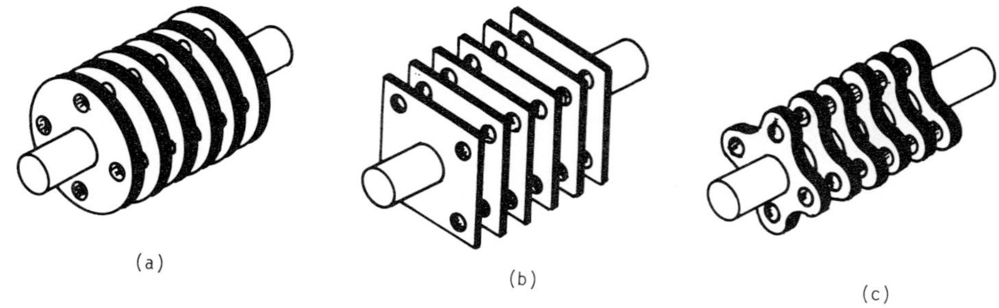

FIGURE 8. Types of hammermill rotors: (a) disk type, (b) plate type, (c) spider type.

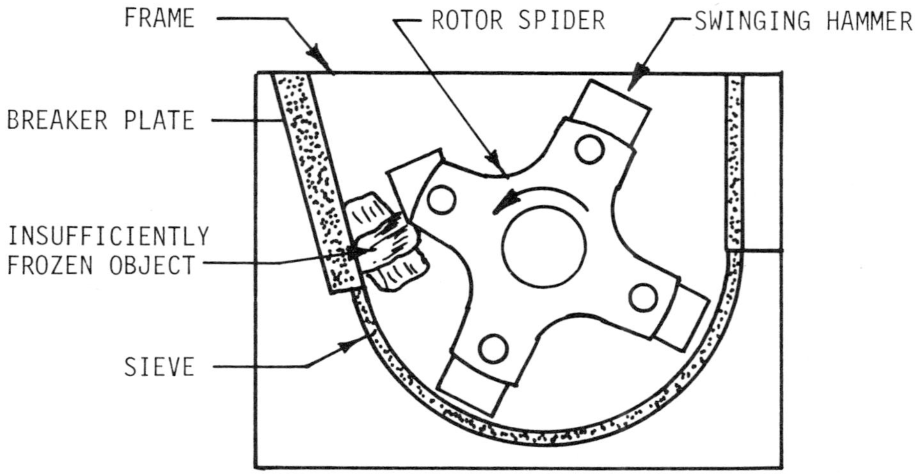

FIGURE 9. A condition in hammermills with spider or plate rotors leading to possible severe damage to the hammermill.

A. Hammers

Various types of hammers are used in hammermills for fragmentizing materials. Figure 10 illustrates the types of swinging hammers which are available. Rigid type hammers are also available for certain rotors; however, with the exception of microgrinding, they are not recommended for cryogenic applications.

The straight and slugger types of hammers are best suited for cryogenic processing. These two types are similar, with the main distinction being in the width of their faces. Straight hammers have narrow faces ranging in width from 3.2 mm to 25 mm (1/8 to 1 in.), while slugger hammers have faces 50 mm (2 in.) in width or wider. Ring hammers can also be used for cryogenic processing; however, they are not as efficient in this type of application as the straight or slugger hammers. (Ring hammers are generally used to pulverize highly friable materials such as coal and sandstone.) Hammer weights range from approximately 85 g (3 oz) each to as much as 341 kg (750 lb) each. The individual weights of the hammers must be considered when the hammers are installed in the rotor. Since there is a normal variation in the weights of the individual hammers, they must be installed in a pattern which will ensure a balanced rotor.

Hammer wear is an important factor in any hammermilling operation whether performed at ambient or cryogenic temperatures. Since the hammers are constantly subjected to impact and abrasive types of wear conditions, the faces of the hammers progressively wear away during operation. When they have worn to a point where they

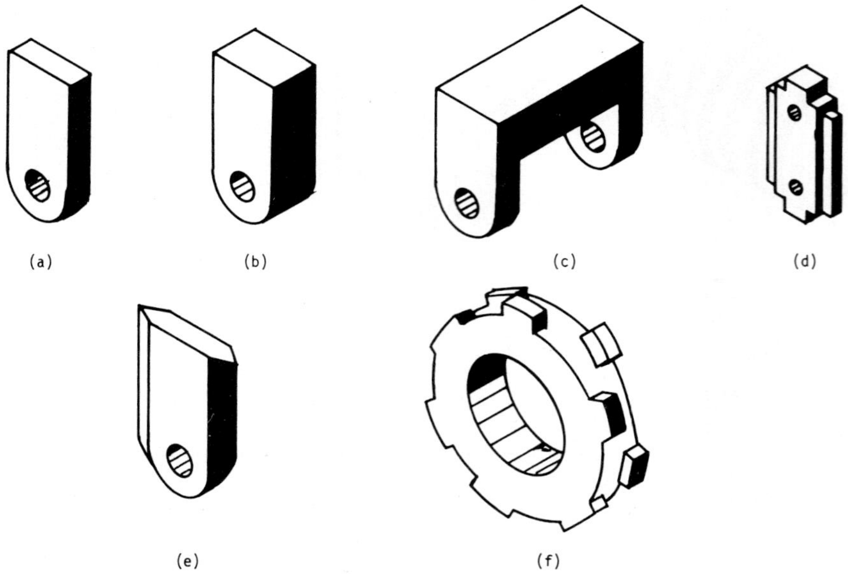

FIGURE 10. Various types of hammers available for hammermills.

are no longer effective, they may be turned to expose a new face or removed for rebuilding. Worn hammers can be reconditioned by building up the worn surfaces with an arc welding process and then applying a hardfacing overlay over the working surfaces (Figure 11). The finished size and weight of the reconditioned hammers must be closely controlled so as to assure a balanced rotor assembly after reinstallation of the hammers. An unbalanced condition in the rotating mass will induce vibrations in the system which will damage the equipment.

B. Product Size Control

Control of the size of the product leaving the hammermill is accomplished by using either perforated metal screens or spaced grate bars located in the discharge portion of the mill (Figure 12). Screens with smaller holes or grate bar sections with narrower openings produce a smaller-sized product. Product size is also influenced by rotor tip velocity and type of hammer used. In cryogenic processing, flat hammers of the straight or slugger type are used; therefore, particle size is controlled by rotor speed and screen size. Figure 13a illustrates the relationship between speed and a given sieve opening. At high speeds the fragmented particles tend to approach the opening at an acute angle to the screen, thus allowing only the smaller particles to pass through. At lower velocities the particles approach at a steeper angle — in effect seeing a larger opening, thus allowing larger particles to pass through. Rotor tip velocities for cryogenic recycling applications range from approximately 36 to 85 m (125 to 300 ft)/sec.

Product size is also influenced by the thickness of the sieves (Figure 13 b and c). For a given speed a thinner sieve produces a larger particle size than does a thicker sieve. The selection of a screen vs. grate bar sieve section should not be determined solely on the basis of the desired final product size but should also take into account the material which is to be processed cryogenically. Grate bar sieves are much stronger than performated metal screens. If processed items contain a considerable fraction of ductile metals, a grate bar sieve should be used in the mill since the grate bar will be better able to withstand the additional processing forces.

For cryogenic processing the coarse fraction opening at the trailing edge of the sieve

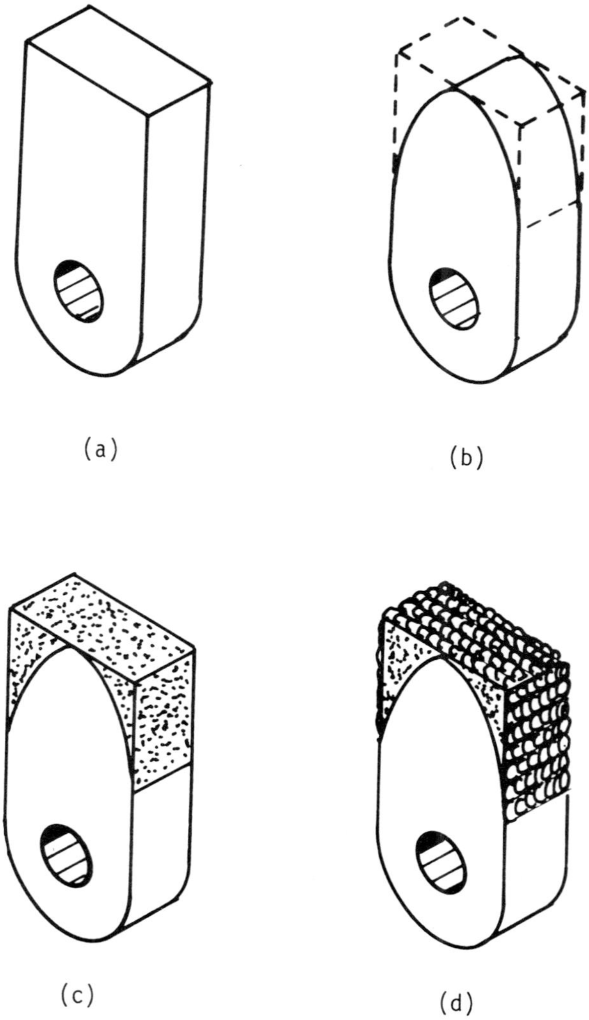

FIGURE 11. Typical wear and reconditioning of hammermill hammers.

should be located below the rotor axis (Figure 14a). This configuration allows gravity to aid in moving the coarse fraction through the openings.

C. Types of Hammermills

Hammermills are available for processing in the down-running or over-running modes, as illustrated in Figure 15. The rotor assemblies and sieve sections of the two types are similar in all respects but in the shape and location of the breaker plates. In the over-running mill, a curved breaker bar section extends from the sieve section to a position near the top of the rotor (Figure 15a). The incoming material strikes the breaker plate a number of times before entering the sieve section; consequently, this type of mill produces a greater percentage of fines in the product. In cryogenic applications, the over-running mills can be used to process items such as scrap tire chips, plastic wares, electrical conductors 5 cm to 15 cm (2- to 6-in. lengths), and small electrical components such as transformers and solenoids.

The down-running mill has a breaker plate extending in a vertical or near vertical

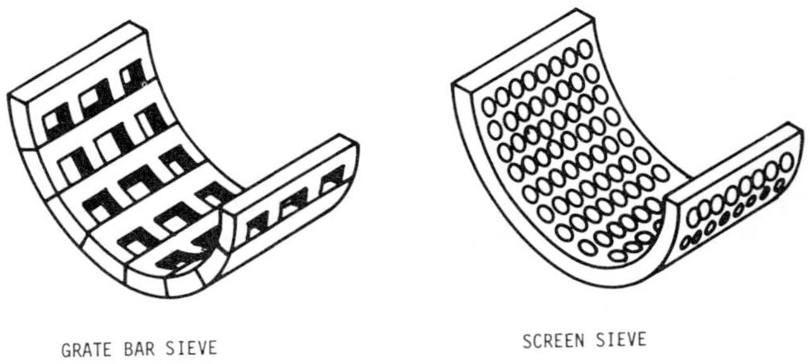

FIGURE 12. Types of sieves used in hammermills.

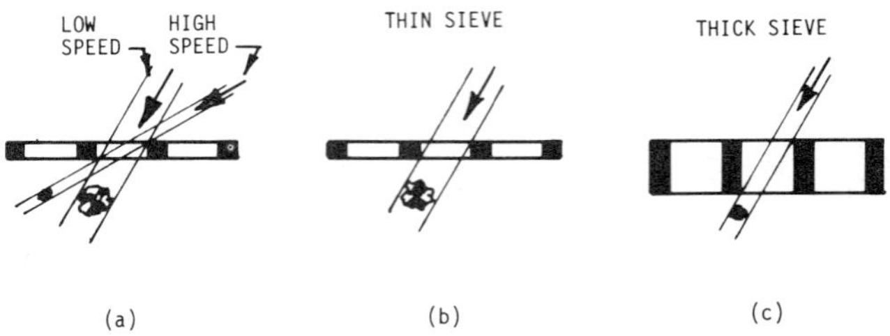

FIGURE 13. Effect of particle velocity and sieve thickness.

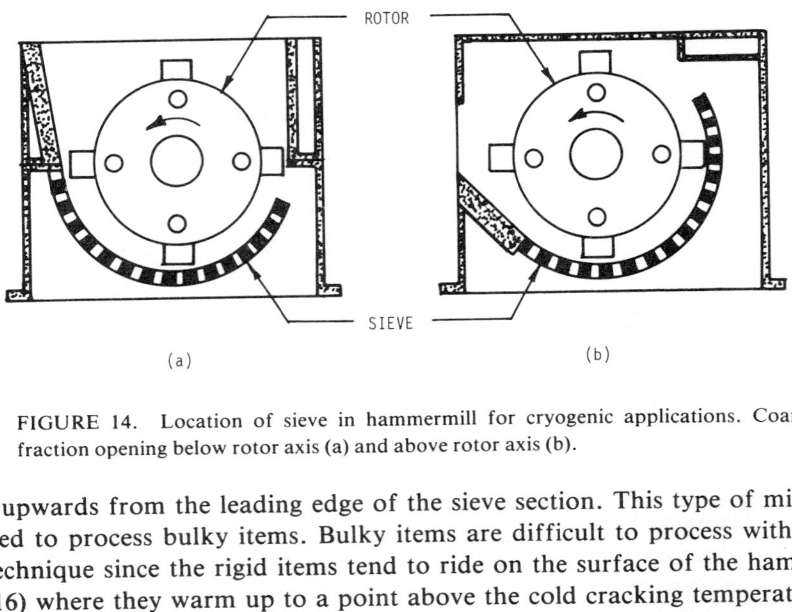

FIGURE 14. Location of sieve in hammermill for cryogenic applications. Coarse fraction opening below rotor axis (a) and above rotor axis (b).

position upwards from the leading edge of the sieve section. This type of mill is generally used to process bulky items. Bulky items are difficult to process with the cryogenic technique since the rigid items tend to ride on the surface of the hammer tips (Figure 16) where they warm up to a point above the cold cracking temperature. The items will then collapse and pass through the mill without being fragmented. For this reason, whole passenger car tires are difficult to process unless a very large hammermill is used. Processing of whole tires is not recommended not only because of the fragmentizing difficulties but also because the bead and belt wires become entangled with the fabric cord material. Separation of the entangled material is not an economical operation at this time. Bulky items such as tires should be preprocessed at ambient temperatures into small pieces which can be efficiently handled by the freezing tunnel. In the case of tires, the fragmented tire chips will yield wire and fabric material in short lengths, which are easily separated.

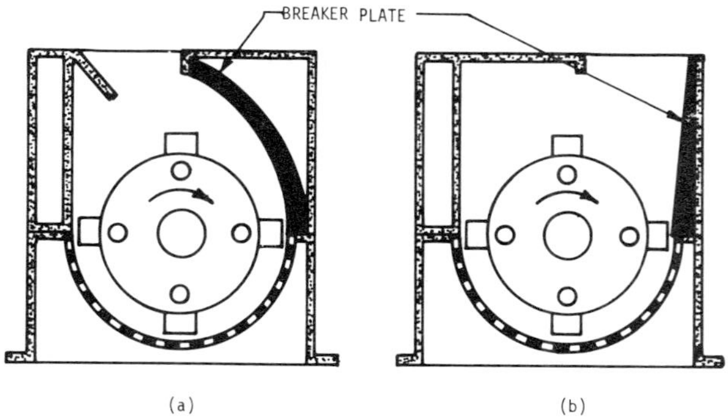

FIGURE 15. Examples of types of hammermills (a) over-running and (b) down-running.

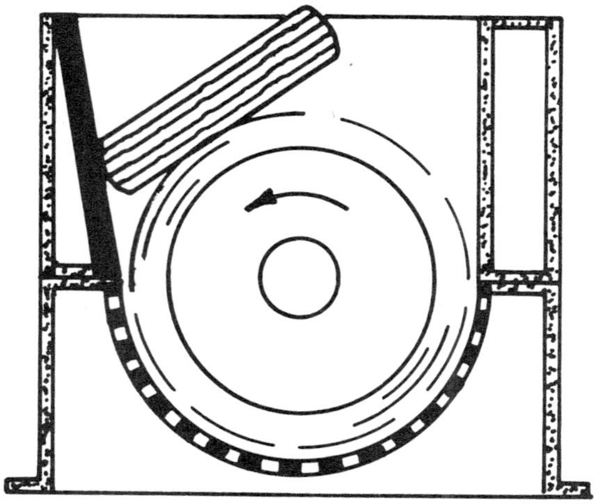

FIGURE 16. A condition which occurs in processing of whole frozen tires.

D. Sieve Modification

The cryogenic processing of items in which the temperatures of all the component materials can be reduced to the cold cracking point creates no special processing problems. All of the materials will fragmentize and pass through sieve in the hammermill. On the other hand, multicomponent items in which one or more of the components remain ductile at cryogenic temperatures do create processing problems. The ductile materials remain in the mill and continue to recirculate through the mill until they are shredded into pieces small enough to pass through the sieve. Copper or aluminum windings from transformers, for example, will be shredded into nuggets or short lengths before they pass through the screen or grate bars. A condition such as this decreases throughput and increases the power requirement.

Processing of scrap tires, either whole or in chip form, creates a similar condition in that after the rubber is fragmented away, the cord material remains in the mill until it is reduced to a fluff which then passes through the sieve. Not only does this require additional power, but it also creates additional handling and separation problems.

To eliminate this condition and allow for the recovery of ductile fractions virtually intact, the sieve section can be modified at its trailing edge to allow most of this material to exit from the working chamber within the first cycle. An unrestricted opening

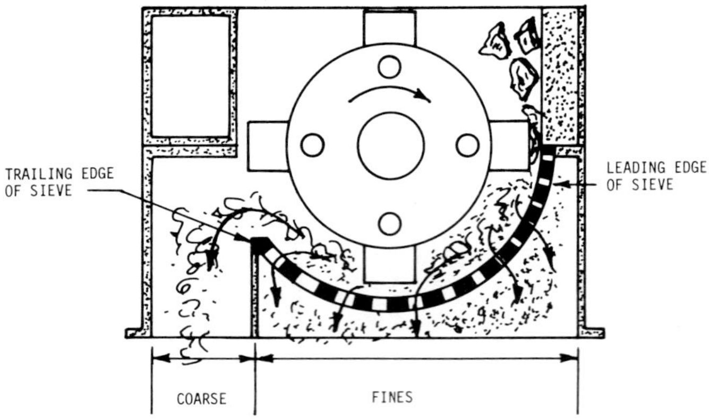

FIGURE 17. Modification to trailing edge of sieve section to permit coarse fraction to leave mill intact.

the full width of the hammermill must be provided at the trailing edge of the sieve section (Figure 17). The length of the opening will depend on the size of the hammermill and the material to be processed. For example, in cryogenic processing of scrap tire chips with a hammermill whose rotor is 1 m (40 in.) wide and rotor diameter is 0.9 m (36 in.), an opening of 0.1 m × 1.0 m (4 in. × 40 in.) at the trailing edge of the grate bars will allow the tire cord fibers and bead wires to leave the mill at the same lengths they entered the mill. As the area of the opening is decreased, more of the fibers recirculate through the mill, resulting in the fraying of the fiber ends. When the opening is completely closed, all of the fibers will be reduced to a fluff. Understandably, the area of the opening must be increased to allow larger components to freely pass through the opening.

In processing electrical transformers 7.5 cm × 7.5 cm × 10 cm (3 in. × 3 in. × 4 in. size) with the hammermill in the previous example, the length of the opening must be increased in size to measure approximately 20 cm (8 in.) to allow the copper windings to leave the mill in an unshredded condition. The ferrous fraction fragments are reduced to approximately 25 mm (1 in.) square pieces and pass through the 50 mm (2 in.) spaces between the grate bars. With this type of modification, it is possible to have two streams of materials leaving the hammermill: a fine fragmentized fraction and a coarse unfragmented fraction.

At the present state of the art with the exception of cryogenic microgrinding equipment the hammermills used for cryogenic processing have been adaptations of equipment manufactured for mining or farming applications. Generally the mining equipment is of much heavier construction than required for cryogenic processing. The heavier construction, however, is required when processing items such as electric motors and larger transformers. Farm-type hammermills are of a lighter construction and can be used for the processing of polymeric material.

In selecting a hammermill for cryogenic applications it is desirable to use a machine with a large rotor diameter since this will provide both a larger breaker bar area and a larger sieve area. Fragmented material should be allowed to pass through the sieves as quickly as possible since retention of the fragmented particles in the mill in the hope of further size reduction only leads to the generation of heat, which degrades the processing operation. Further size reduction of the product must be accomplished in a microgrinding operation where a cryogen can be injected into the grinding chamber to keep the product at a reduced temperature. Screen openings and grate bar spacings which are too small during first-stage cryogenic processing lead to a longer dwell time

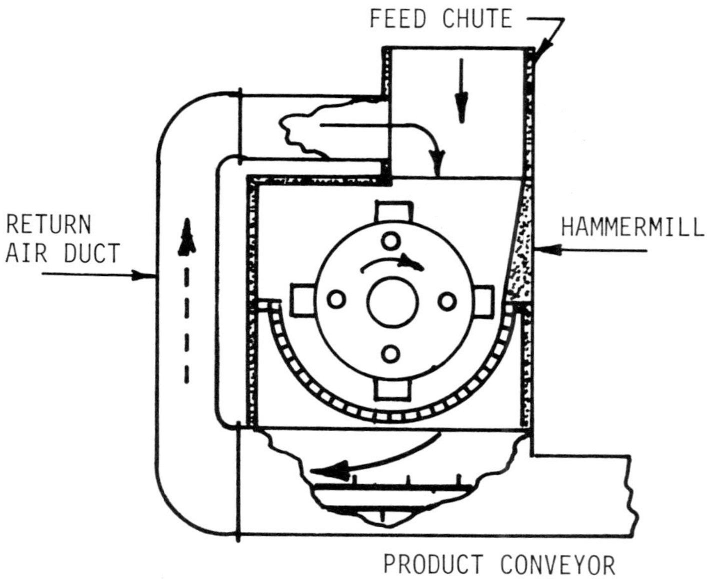

FIGURE 18. Recirculating air system for dust control.

of the product in the mill resulting in heat build-up. Under these conditions, polymeric materials begin to smear along the sieve leading to complete blinding of the sieves.

E. Materials for Hammermill Construction

Conventional hammermills which are used at ambient temperatures are fabricated from carbon and manganese steels. Generally, the frame assemblies are fabricated from carbon steel while those parts which are subjected to impact loading conditions are fabricated of manganese steel. Manganese steels are used to fabricate parts such as breaker plates, hammers, and liner or wear plates. This type of construction is adequate for the equipment used in first-stage cryogenic processing. Since cryogens are not injected into the hammermilling chamber, the temperatures of the mill components are not reduced to levels where they are susceptible to brittle failure.

Microgrinding equipment, however, may be operated with the injection of a cryogen into the grinding chamber. Equipment for this type of application is constructed of stainless steel which remains ductile at reduced temperatures.

F. Dust Control Equipment

Hammermills act as fans and consequently generate a considerable airflow through the sieve section. In cryogenic processing (especially of polymeric materials), a considerable quantity of fines are generated which can be lost to the atmosphere. To prevent contamination of the atmosphere and also to recover this valuable material, a number of dust-control systems are available for hammermills. The two basic dust-control systems are the recirculating air and the cyclone collector types.

1. Recirculating Air System

The recirculating air system is the simplest dust-control system. In this system, the hammermill draws the air discharged through the sieve section back to the feed opening in essentially a closed loop system (Figure 18). A system of this type is usable for processing metal items such as starters, generators, or electric motors. This sytem only minimizes the amount of dust lost to the atmosphere and should not be used where total dust control is required.

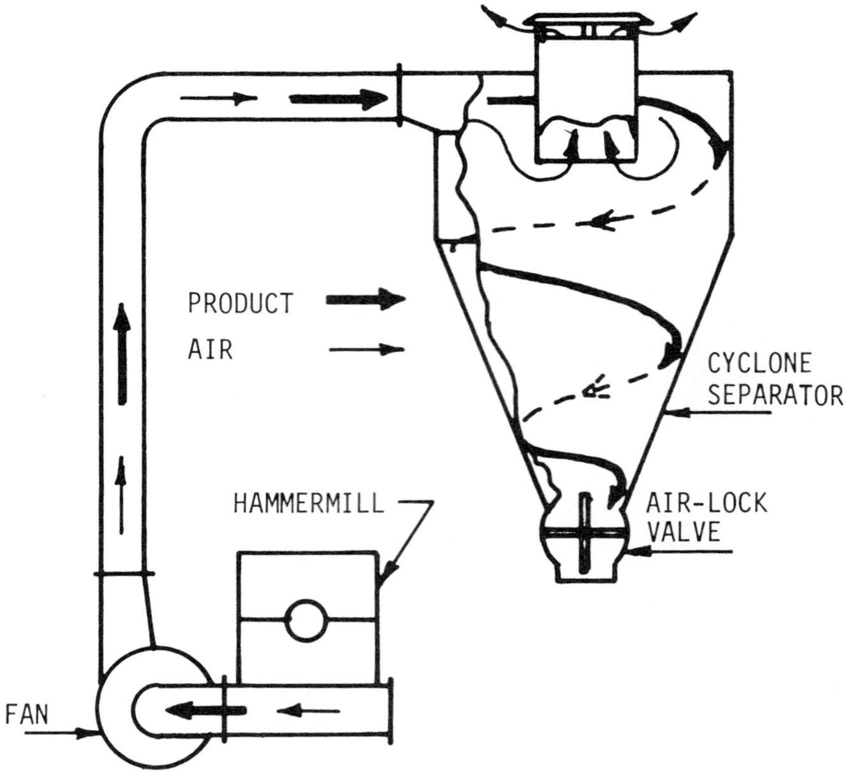

FIGURE 19. Pressure type air system.

2. Cyclone Collectors

Cyclone collectors are able to provide total dust control in cryogenic processing, especially polymeric materials. The three types of cyclone collectors are the pressure air, return air, and vacuum air systems. These systems can be used only for processed materials which are air-transportable. An auxiliary fan pulls the processed material from the mill and blows it into the chamber of the cyclone collector. The product follows a path along the inside walls of the cone and progresses downward toward the product discharge port as its tangenital velocity decreases (Figure 19). The air which contains fine particulate matter is discharged through the central chimney. This is the most common system of dust control for hammermills.

3. Return and Vacuum Systems

When it is necessary to retain the particulate matter which is normally lost in the pressure air system, a return air or a vacuum air system can be used (Figures 20 and 21). The basic operation of both systems is similar in that the product is separated from the air stream in a cyclone collector, and the fine particulate matter is collected in an auxiliary cloth collector system. In both systems, the major portion of the air discharged from the cyclone collector is returned to the hammermill.

Although the initial investment of a vacuum system is higher than for the return air system, it does have two main advantages. Since most of the system is maintained at a negative pressure, fine material is not lost through leaks in the system. It is, therefore, possible to provide absolutely dustless cryogenic processing. Negative pressure is maintained in the product lines by placing the fan downstream from the cyclone; conse-

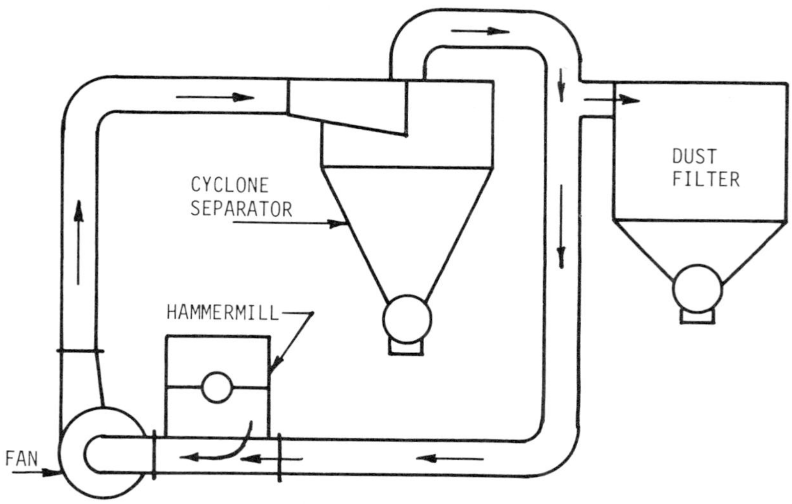

FIGURE 20. Return air system.

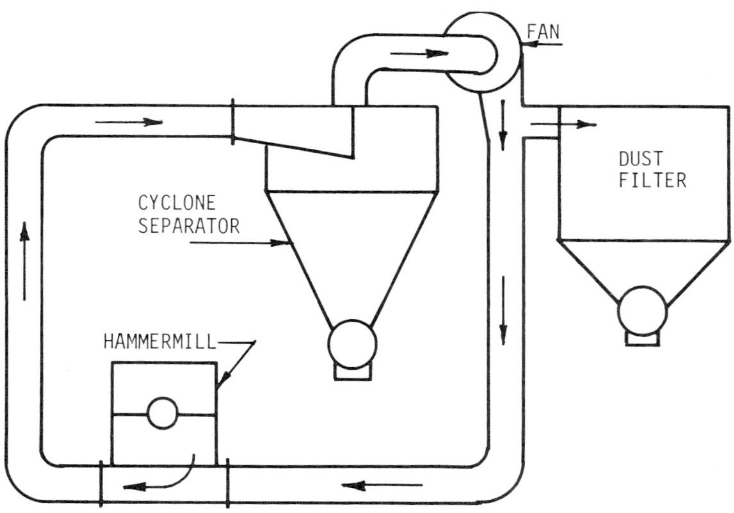

FIGURE 21. Vacuum air system.

quently, only fine material passes through the fan blades. Fan blade life is thereby extended resulting in a lower maintenance cost of the system.

The vacuum air system is highly recommended for the processing of scrap tires and plastics since recovery of virtually all the fines is assured.

G. Product Separation

After scrap materials have been pulverized by the cryogenic technique, the resulting product must be processed further to separate the various fractions. The separation of these fractions is accomplished by using methods such as magnetic attraction, oscillating or rotating screens, gravity systems, or fluidized bed techniques.

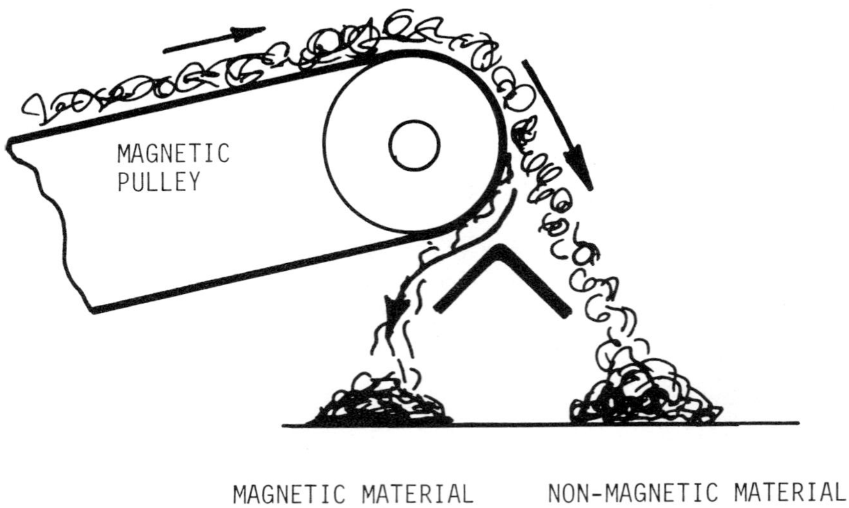

FIGURE 22. Principle of operation of magnetic head pulley.

1. Magnetic Separation

Magnetic fractions can be removed from a product stream by using suspended separation magnets or magnetic pulleys or drums. Permanent or electromagnets in these systems attract only magnetic fractions such as iron and nickel to themselves. With these systems, a product stream can be split into magnetic and nonmagnetic streams.

a. Magnetic Pulleys

Permanent magnet or electromagnetic head pulleys can be installed in a product conveyor system to remove any magnetic fractions which are mixed with the product. Figure 22 illustrates the operation of a magnetic head pulley. As the product begins to move over the pulley, the magnetic material is attracted and held to the belt by the influence of a magnetic field. After traveling to the bottom of the pulley, the magnetic material falls away from the belt. The nonmagnetic materials fall away from the pulley in a normal trajectory. Magnetic pulleys are available in 0.3 m to 1.5 m (12 to 60 in.) diameters and belt widths.

There are two limitations in the selection of magnetic pulleys: belt velocity and burden height. Belt velocities above 152 m/min (500 ft/min) and burden depths greater than 20 cm (8 in.) should not be used. These maximum limits apply to the 1.5 m (60 in.) diameter pulleys. Progressively lower values for belt velocity and burden height must be used as the pulley diameter decreases. For example, the maximum values for a 0.3 m (12 in.) diameter pulley would be 53 m/min (175 ft/min) for the belt velocity and 7.5 cm (3 in.) for the burden height. Operating magnetic pulleys above the recommended values increases the likelihood of retaining some of the ferrous fraction in the nonmagnetic stream.

The choice of permanent or electromagnetic pulleys in diameters less than about 0.75 m (30 in) is a matter of convenience. Permanent magnets require no additional power supply as is required for an electromagnet. The electromagnetic pulley, however, can be easily turned on or off. Electromagnetic pulleys are used for diameters greater than 0.75 m (30 in.) because of the greater magnetic strength requirements.

b. Magnetic Drums

Figure 23 illustrates the separation of ferrous materials with a magnetic drum oper-

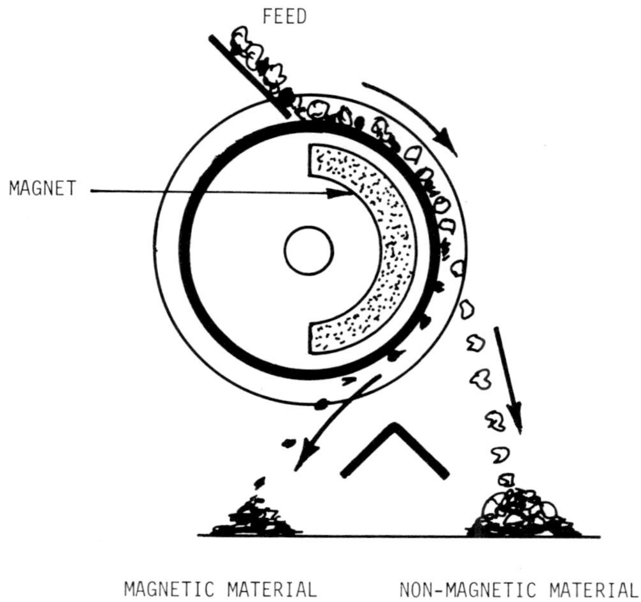

FIGURE 23. Magnetic drum operating in overfeed mode.

ating in the overfeed mode. A permanent or electromagnet remains in a fixed orientation inside of a rotating drum. The position of the magnet can be changed through 360° of rotation in order to change the pickup and drop-off points on the drum. Operation of the system is similar to that of a magnetic head pulley.

As material is fed onto the rotating drum, the magnetic field holds the ferrous material onto the drum until it is carried to the drop-off point below the pulley. Nonmagnetic material follows a normal trajectory from the rotating drum.

Illustrated in Figure 24 is the operation of a magnetic drum in the underfeed mode. Material is fed to the drum near the four- or five-o'clock position where the magnetic fraction is picked up and carried over the top of the drum to the drop-off point.

Magnetic drums are available in diameters ranging from about 0.3 m to 1.5 m (12 to 60 in.) and face widths from 0.2 m to 1.72 m (8 to 68 in.). Operating surface velocity limitations are similar to those for magnetic head pulleys. Surface velocities for operation in the overfeed mode are generally about 30% lower than for the underfeed mode. For example, a nominal operating surface velocity for a 1.5 m (60 in.) diameter drum would be about 143 m/min (470 ft/min) in the underfeed mode and about 100 m/min (330 ft/min) in the overfeed mode.

c. Suspended Separation Magnets

Suspended separation magnets are devices suspended over the product from which the ferrous fraction must be removed. Figure 25 illustrates a stationary suspended system. This system is not recommended for cryogenic processing.

The self-cleaning system as illustrated in Figure 26 is recommended for continuous processing operations. During operation, the material is attracted to the magnet in the center section of the unit. The moving cleated belt carries the material to one side, where it is discharged. These units can be positioned transverse to the product flow (Figure 26b). Self-cleaning suspension magnets are available with either permanent or electromagnets. These systems can be used with either belt conveyor or shaker table product lines. When used in conjunction with shaker tables, the table below the suspended magnet must be made of a nonmagnetic material.

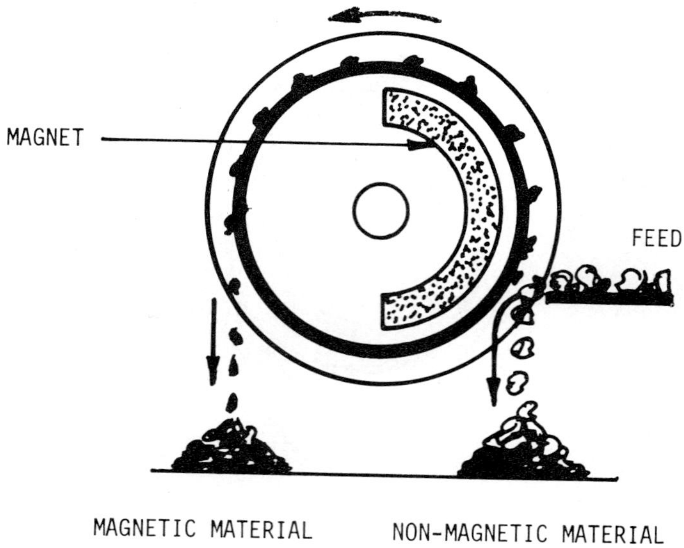

FIGURE 24. Magnetic drum operating in underfeed mode.

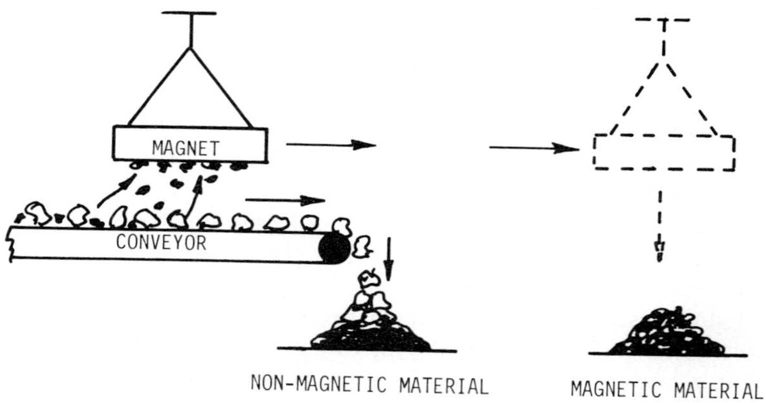

FIGURE 25. Operating principles of stationary overhead magnet.

2. Screen Separation

The simplest method for separating the nonmagnetic split of processed material into several fractions is with an air-screen system. A system of this type could be used, for example, to separate the rubber crumb from the fabric cord in the cryogenically processed scrap tire product.

Although there are many variations of these systems, the basic operating principles are the same. The top scalping screen (Figure 27) removes the larger material from the incoming product while the lower screens refine and size the remaining material. Eccentric shaft, eccentric flywheel, or pitman drives oscillate the screens to move the materials across the screens while an air stream is directed through the screens to aid in the cleaning action. Equipment of this type is available with throughputs up to 54 metric tons (60 tons)/hr.

3. Vacuum Gravity Separators

Vacuum gravity systems can provide a complete spectrum of separation ranging from the very lightest to the heaviest, based on their specific gravities. Figure 28 illus-

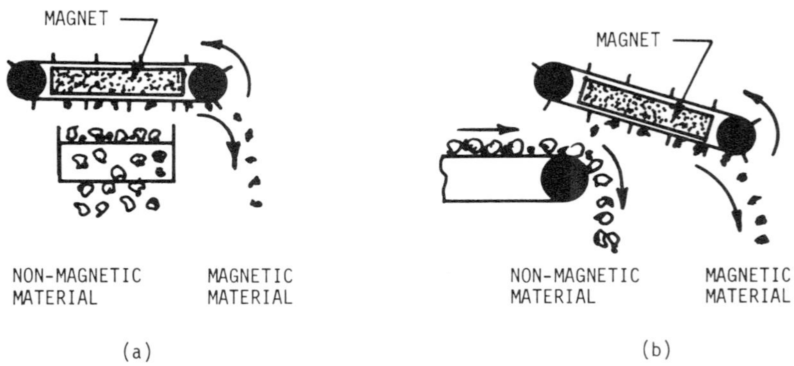

FIGURE 26. Continuously self-cleaning suspended magnet separators.

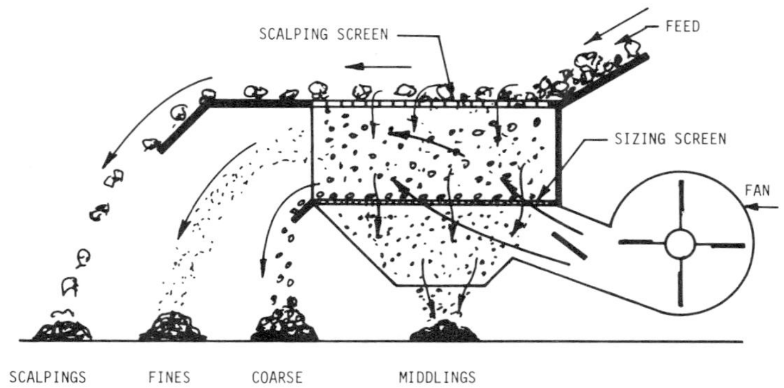

FIGURE 27. Simplified schematic of an air-screen separator.

trates the operation of a typical vacuum gravity system. After entering the screen section at the lower portion, the product moves upwards along the oscillating screen. Air in the closed loop system is pulled through the product to lift off the fines and carry them to a dust bag recovery system. The various fractions, culls, lights, mediums, and heavies are split from the stream at appropriate locations along the screen. The heaviest fraction is recovered from the highest position on the screen while the cull fraction which is the lightest is recovered from the lowest position.

The various fractions of cryogenically processed materials such as insulated electrical conductors, solenoids, relays, or scrap tire chips can be separated in a vacuum gravity system. Since these systems are closed loop systems, they provide total air and dust control, which results in no building heat losses or pressure built-up. The 2.7 tonne (3 tons)/hr processing capacities of the gravity separators are considerably lower than those obtainable with the air-screen systems.

4. Other Separation Techniques

Several of the newer techniques being worked on for the separation of processed scrap materials are the liquid density, ferrofluid, and eddy current systems.

a. Liquid Density Systems

Mixed plastics can be split into two fractions by placing the mixture into a liquid bath of a specific density. Plastics with densities lower than that of the liquid will float while those with higher densities will sink in the bath. By changing the density of the

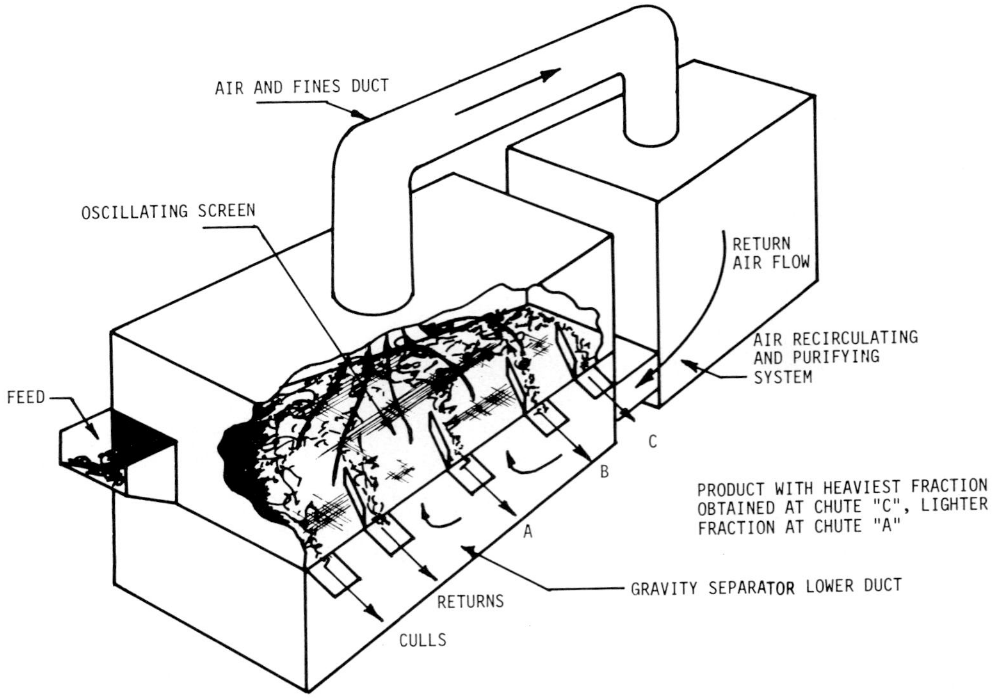

FIGURE 28. Schematic of a gravity system separator.

fluid, the light and heavy fractions can each be split again into float and sink fractions. This procedure can be continued until all the component plastics have been separated out.

b. Ferrofluid Technique

The ferrofluid technique operates on the same principle as applied to the liquid density system. In this system, the density of the ferrofluid is varied by changing the strength of the magnetic field passing through the fluid. This system effectively separates nonmagnetic materials into float and sink fractions.

c. Eddy Current Technique

The eddy current system for the separation of nonmagnetic scrap from other solid wastes operates on the principle that eddy currents induce electrical currents in metal particles resulting in a force on these particles. One such system has a series of magnetic strips of alternating polarity placed at an angle of about 45° to the product flow. As the nonmagnetic metal flows down an inclined ramp past the magnets, the alternating magnetic field generates a force in the metal particle causing it to be deflected away from the main product stream.

V. TYPICAL CRYOGENIC SCRAP PROCESSING PLANT

The type of equipment discussed in the preceding sections is adaptable to setting up a cryogenic scrap processing plant. Illustrated in Figure 29 is a typical processing plant for fragmenting scrap tire chips and then separating the product into three fractions: rubber crumb, wire, and fabric cord. The processing cycle begins with the shredding of scrap automotive tires into uniform-sized chips. A uniform chip size not only in-

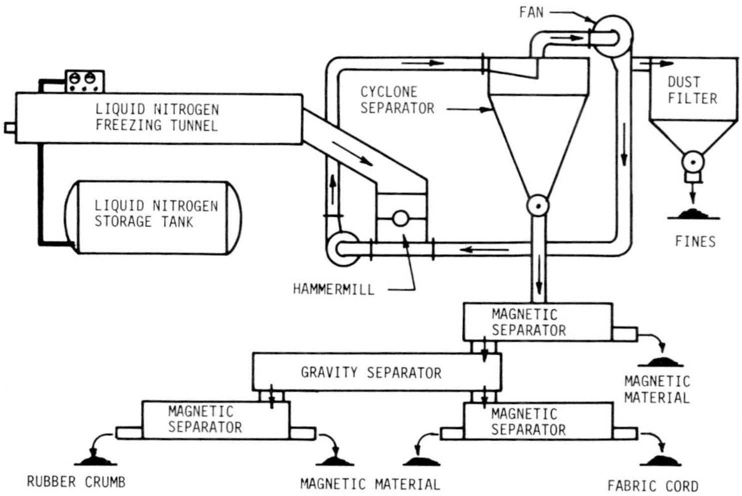

FIGURE 29. Schematic of a cryogenic plant for processing scrap automobile tire chips.

creases freezing efficiency but provides a more uniform particle size of the fragmented tire chips. The smaller tire chips can be fragmented with a lighter-duty hammermill.

The tire chips are fed into the freezing tunnel in a thin uniform layer. As the chips progress through the tunnel, their temperature is constantly being reduced, initially by liquid nitrogen vapors and finally with a liquid nitrogen spray. At this point, the surface layer of the product is supercooled; consequently, an equilibration period is provided downstream of the spray section to allow the temperature to stabilize at −68°C (−90°F) throughout the entire mass of the tire chips.

Liquid nitrogen is drawn from a cryogenic storage vessel located near the processing site and is delivered to the freezing tunnel via an insulated piping system.

The size of the freezing tunnel system is governed by the required throughput and the system's heat extraction capability $kJ/kg/cm^2$ ($Btu/lb \cdot ft^2$). Assuming about a 13.5 metric tons (15 tons)/hr throughput of tire chips through a 1.3 m (50 in.) wide freezer, the tunnel would be approximately 24 m (80 ft) in length. In some cases, throughput can be pushed to about 140% of normal; however, the per-pound freezing cost increases because the cryogen is not used efficiently in this mode of operation.

Frozen tire chips move directly from the freezer into the hammermill where the rubber is fragmented away from the cord and wire in a matter of seconds. The mixture of rubber crumb, cord, and wire is then air-carried to the cyclone vacuum air separator. An air lock valve assembly at the product discharge part of the separator assumes proper air balance in the air system. The auxiliary bag collector system separates the fines from the airstream and collects them in the enclosed hopper. The fines can be removed from this hopper during processing if an air lock valve is used.

The first stage of separation for the product flowing from the cyclone is magnetic separation where the bead and steel belt wires are removed. The nonmagnetic fraction from this stage is then split into cord and rubber fractions in a vacuum gravity separator.

Magnetic separation is used for each of the splits to remove any wires which were missed in the first magnetic seapration stage.

If a fine rubber dust is the desired product, it is advantageous to microgrind the rubber crumb immediately after separation since the rubber particles are still at a reduced temperature.

VI. CONCLUSION

Cryogenic processing of scrap materials for the reclaiming of various fractions is still in the infancy stage. It is being accomplished by using modified freezers and hammermills which were designed for other purposes. Sanitation requirements for food handling equipment makes it mandatory that food freezing units be fabricated of stainless steel with polished surfaces and welded joints. A considerable cost savings can be realized by using lower cost materials for all surfaces not exposed to the cryogenic environment.

Presently there is no fragmenting equipment available which is specifically designed for cryogenic processing of scrap materials such as scrap tires. Cryo-microgrinding is a well-established technology; however, microfragmenting equipment simply cannot be scaled upwards to meet the requirements for scrap processing. Fragmentizers designed for rock or coal crushing also do not meet the requirements for cryogenic processing without modifications. This comminution equipment is designed to fragment brittle materials; however, the processing is carried out at room temperatures and the product remains brittle irrespective of its dwell time in the hammermill. In cryogenic processing the scrap material must be fragmented into fine granular material within one or two cycles through the mill. A prolonged dwell time in the mill allows the material to increase its temperature above the cold cracking point, preventing additional fragmenting. A hammermill for cryogenic processing of multicomponent materials such as scrap tires should have a large diameter rotor which permits the use of a long sieve section along with a discharge opening at its trailing edge. With this arrangement the dwell time for both the brittle and ductile fractions is kept to a minimum, reducing the likelihood of overheating the scrap material.

The equipment downstream of the hammermill creates no significant problems allowing the use of standard "off the shelf" equipment.

REFERENCES

1. Airco Cryogenics, Kryoaire, Bull. 1029, Airco Cryogenics, Irvine, Calif.
1a. Airco Cryogenics, Kwik Freeze Systems, Airco Cryogenics, Irvine, Calif.
2. Air Products and Chemicals, The Cryo-Quick Freezer, Air Products and Chemicals, Allentown, Pa.
2a. Air Products and Chemicals, Cryo-Grind Systems, Air Products and Chemicals, Allentown, Pa.
3. Automotive-Industrial Marketing Corporation, Tiregon, Automotive-Industrial Marketing Corp., Portland, Ore., 1973.
4. Cryogenic Recycling International, Inc., Mobile Material Chipping Plant, Cryogenic Recycling International, Inc., La Crosse, Wis.
5. Cumberland Engineering Co., Inc., Cumberland Granulating Machines, Bull. 73-G, Cumberland Engineering Co., Inc., Providence, R.I., 1971.
6. Dings Magnetic Co., Magnetic Pulleys, Bull. C-2000C, Dings Magnetic Co., Milwaukee, Wis., 1969.
7. Dings Magnetic Co., Ceromox (DFC) Dry Drum Magnet, Bull. 1125, Dings Magnetic Co., Milwaukee, Wis.
8. Fitzpatrick Company, Comminution, Bull. INF-62, Fitzpatrick Co., Elmhurst, Ill., 1969.
9. Forsberg, Inc., Screen-Aires; Bull. 770-1, Forsburg, Inc., Thief River Falls, Minn.
10. Forsburg, Inc., Vacuum Gravity Separator, Bull. 773-2, Forsburg, Inc., Thief River Falls, Minn.
11. Gruendler Crusher and Pulverizer Co., Industrial Reduction Hammermills, Gruendler Crusher and Pulverizer Co., St. Louis, Mo.
12. Gruendler Crusher and Pulverizer Co., Many a Plant Has Ground to Success with a Gruendler, Cat. No. M-527-C, Gruendler Crusher and Pulverizer Co., St. Louis, Mo.

13. Hammermills, Inc., Bulldog Impact Type Hammermills, Bull. HAM-306-67, Hammermills, Inc., Cedar Rapids, Iowa.
14. Hammermills, Inc., Bulldog Refuse Shredders, Bull. HAM-703, Hammermills, Inc., Cedar Rapids, Iowa.
15. Iowa Manufacturing Co., Commander Series Tandem Crushing and Screening Units, Bulletin COMS-2, Iowa Manufacturing Co., Cedar Rapids, Iowa.
16. Iowa Manufacturing Co., Hammermills, Bull. HAM-4, Iowa Manufacturing Co., Cedar Rapids, Iowa.
17. Iowa Manufacturing Co., Limemills, Bull. LM-1, Iowa Manufacturing Co., Cedar Rapids, Iowa.
18. Iowa Manufacturing Co., Utilized Portable Secondary Crushing and Screening Plants, Bull. Unit-8, Iowa Manufacturing Co., Cedar Rapids, Iowa.
19. Integral Process Systems, Inc., Cryotransfer Series Liquid Nitrogen Flash Freezing Systems for Food Freezing, Form. No. AJ44, Integral Process Systems, Inc., Park Forest, Ill.
20. Jeffrey Manufacturing Co., Type AB and WB Hammermills and Wood-Hogs, Bull. 1129, Jeffrey Manufacturing Co., Columbus, Ohio.
21. Liquid Carbonic Corporation, Cryotransfer XB 402 Liquid Nitrogen Flash Freezing System, Form No. 6307, Liquid Carbonic Corp., Chicago, Ill., 1977. Liquid Carbonic Corporation, Cryogenic Grinding Systems, Form 6625, Liquid Carbonic Corp., Chicago, Ill.
22. Logemann Brothers Co., Scrap Metal Guillotine Shears, Cat. GS 272, Logeman Brothers Co., Milwaukee, Wis.
23. Metropolitan Disposal Corporation, Tire Hawg, Metropolitan Disposal Corp., Portland, Ore.
24. Ohio Magnetics, Ohio Scrap Handling Magnets and Auxiliary Equipment, Bull. OM 101, Ohio Magnetics, Maple Heights, Ohio, 1972.
25. Pennsylvania Crusher Corporation, Material Reduction Equipment, Bull. 4022, Pennsylvania Crusher Corp., Broomall, Pa., 1974.
26. Portec, Inc., Pioneer Impact Crushers and Hammermills, Form 718-A, Portec Inc., Minneapolis, Minn.
27. Siso A/S, Siso Rasper, Siso A/S, Copenhagen, Denmark.
28. Stearns Magnetics, Inc., Electromagnetic Pulleys, Bull. 1013, Stearns Magnetics, Inc., Cudahy, Wis., 1973.
29. Stearns Magnetics, Inc., Permanent Suspended Separation Magnets, Bull. 1036, Stearns Magnetics, Inc., Cudahy, Wis., 1972.
30. Stearns Magnetics, Inc., Electromagnetic Type "LD" Drum, Bull. 1042, Stearns Magnetics, Inc., Cudahy, Wis., 1970.
31. Sturtevent Mill Co., Sturtevant Whirlwind Centrifugal Separator, Bull. 087, Sturtevent Mill Co., Boston, Mass.
32. W. C. Wiedenmann and Son, Inc., Model LF Filtramatic Dust Filter, Bull. F400, W. C. Wiedenmann and Son, Kansas City, Mo.
33. W. C. Wiedenmann and Son, Inc., Rotary Air Lock Valves and Feeders, Bull. V-12, W. C. Wiedenmann and Son, Inc., Kansas City, Mo.
34. W. C. Wiedenmann and Son, Inc., H. P. Cyclone Dust Separators, Bull. HC-11, W. C. Wiedenmann and Son, Inc., Kansas City, Mo.
35. Williams Patent Crusher and Pulverizer Co., Inc., Type "C" Crushers, Shredders, Hogs, and Impactors, Bull. 784, Williams Patent Crusher and Pulverizer Co., St. Louis, Mo.
36. Williams Patent Crusher and Pulverizer Co., Metal Turnings Crushers, Bull. 799, Williams Patent Crusher and Pulverizer Co., St. Louis, Mo.
37. Williams Patent Crusher and Pulverizer Co., Hammermills for Industry, Bull. 776, Williams Patent Crusher and Pulverizer Co., St. Louis, Mo.
38. Williams Patent Crusher and Pulverizer Co., Air Separators, Bull. 774, Williams Patent Crusher and Pulverizer Co., St. Louis, Mo.

Chapter 7

CRYOGENIC STORAGE VESSELS

J. M. Venetucci

TABLE OF CONTENTS

I. INTRODUCTION

The shape of the storage vessel for cryogens can widely vary. However, the majority of commercial vessels used for industrial or medical applications are cylindrical in shape with dished heads used for the end walls. The vessels are available as horizontal or vertical designs. The vertical cryogenic vessel appears to have gained wider acceptance because it requires less installation area as compared to the horizontal vessel of the same capacity. Low-profile horizontal vessels, however, are sometimes preferred if sufficient area is available and a strong desire exists to preserve the appearance of a building or surrounding area. The availability of spherical type vessels for typical industrial installations is decreasing, which may be caused by the possible higher cost of fabrication. In comparison, however, extremely large spherical vessels with storage capacities of 851,692 ℓ (225,000 gal) or more are used as "holding" vessels at air-separation plants. Figure 1 shows typical holding spheres at a plant location. These vessels would be the initial storage vessels from which the cryogen is transferred to specially designed, mobile cryogenic transport trailers or rail cars for delivery to other storage vessels at consumer locations. A typical vacuum-insulated cryogenic trailer is shown in Figure 2.

Cryogenic vessels can be generally classified in one of the following insulated categories:

1. High vacuum
2. High vacuum with powder or fiber-filled
3. High vacuum with multiple radiation shields
4. Rigid cellular insulation

The high vacuum-insulated vessels are generally referred to as Dewars, named after James Dewar, who invented the vacuum-insulated container for liquefied gases in 1892. This unique design method still remains as the best-known basic method to store cryogenic liquids. New techniques, developments, and advances on the original design concept have been directed toward the reduction of heat transfer to the inner container. Some improvements have been attained with the use of powder or multiple layers of insulation in the vacuum space.

The following sections discuss the storage and safe handling of cryogenic liquid.

II. CRYOGENIC STORAGE VESSELS

A cryogenic storage vessel is a receiver which has been designed to efficiently hold and dispense a cryogen as required. The basic vessel design must include components to ensure the safe operation of a pressure vessel at extremely low temperatures. The operation of the various components are described later. Most vessels used for industrial installations are designed and fabricated based on the American Society of Mechanical Engineers Boiler and Pressure Vessel Code, Section VIII. These suggested code rules, if they are followed, are designed to ensure the safe operation of the pressure vessel.

The storage container usually will consist of two vessels, one surrounding the other. A typical vessel schematic is shown in Figure 3. The inner vessel is designed as the cryogenic liquid pressure container. Inner vessels are constructed of materials such as stainless steel, aluminum, or other materials capable of withstanding low cryogenic temperatures. An outer vessel which is not fabricated as a pressure vessel is used to protect the inner vessel and provide an insulated space between the two vessels.

FIGURE 1. Typical holding spheres at a plant location. (Courtesy of Liquid Carbonic Corporation, Chicago.)

FIGURE 2. Typical vacuum insulated cryogenic trailer. (Courtesy of Liquid Carbonic Croporation, Chicago.)

The inner vessel is supported by a suspension system designed to transfer a minimum amount of heat and still be flexible enough to permit contraction due to extreme temperature change. The annular space between the two vessels is filled with insulating material and contains a vacuum to reduce heat transfer to the inner vessel. The insulating material can be of a powder type commonly referred to as "perlite", fiber type, or a combination of fiber and multiple radiation shields.

The capacity of a vessel is usually established by the nominal amount of liquid that it contains. In addition to the liquid volume, there is generally a vapor space allowance above the liquid. This vapor space is excluded from the total liquid volume of the vessel when defining the size of the vessel. For example, a 37,745-ℓ (10,500-gal) vessel would contain 37,853 ℓ (10,000 gal) of liquid at the normal maximum level and 39,745 ℓ (10,500 gal) of liquid when completely filled.

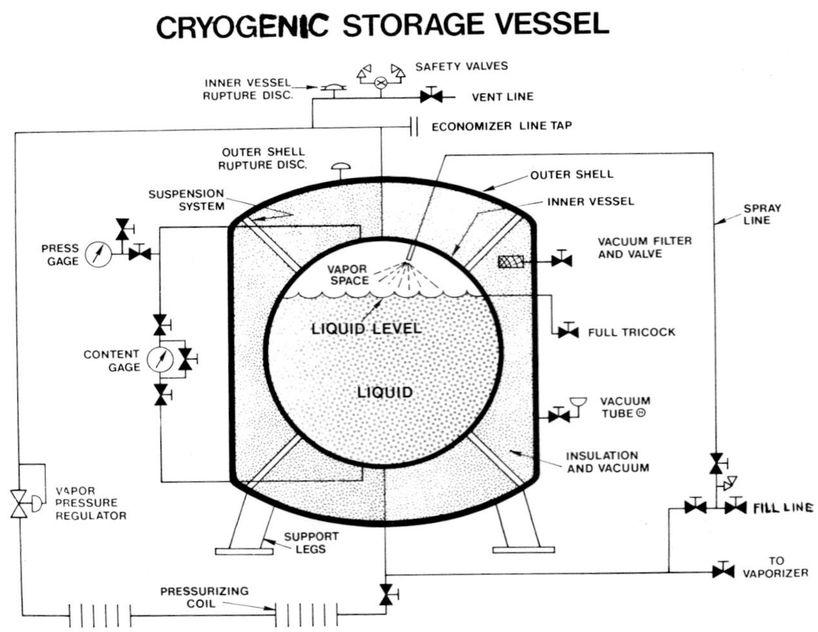

FIGURE 3. Schematic of a typical cryogenic storage vessel. (Courtesy of Liquid Carbonic Corporation, Chicago.)

The generally accepted standard of providing a 10% vapor space in a vessel allows for adequate filling as well as preventing the liquid from being carried through the vent piping. Many typical vessels have a full trycock line which is installed at the location of the maximum liquid level and the 10% vapor space. During filling operations, the full trycock line is used to determine when the vessel is full. When a small stream of liquid appears at the outlet of the full trycock line, the vessel is at its maximum full liquid level.

A. Small Containers and Portable Containers

Generally speaking, the small volume containers which are used within a laboratory are called portable Dewars. These containers vary in size and can have large open tops or small discharge openings from which the cryogenic fluid can be poured. Being at atmospheric pressure, a loose-fitting lid is used to cover the opening of the Dewar. Figure 4 illustrates the construction of a typical portable Dewar. Figure 5 shows the outside protective cover of the Dewar with the internal spheres hidden from view.

Larger portable containers which contain up to 1135 ℓ (300 gal) are available, fabricated to withstand punishment during transportation. As the containers become larger in volume, they generally are capable of storing the cryogenic liquids for a longer period of time. Also, as their capacity increases, additional features are incorporated into them to make them more versatile.

Figure 6 shows the construction view of a larger pressure-type portable Dewar. This type of container is a pressure vessel and is equipped with integral piping components which allow for the convenient and easy dispensing of the cryogen. The unit is also equipped with both a pressurizing and vaporizing coil. These coils provide the means to withdraw large and continuous amounts of liquid or gaseous products. Figure 7 shows a typical 200-ℓ (53-gal) portable container complete with a mobile cart.

When larger amounts of cryogens are required, the use of a mobile portable vessel

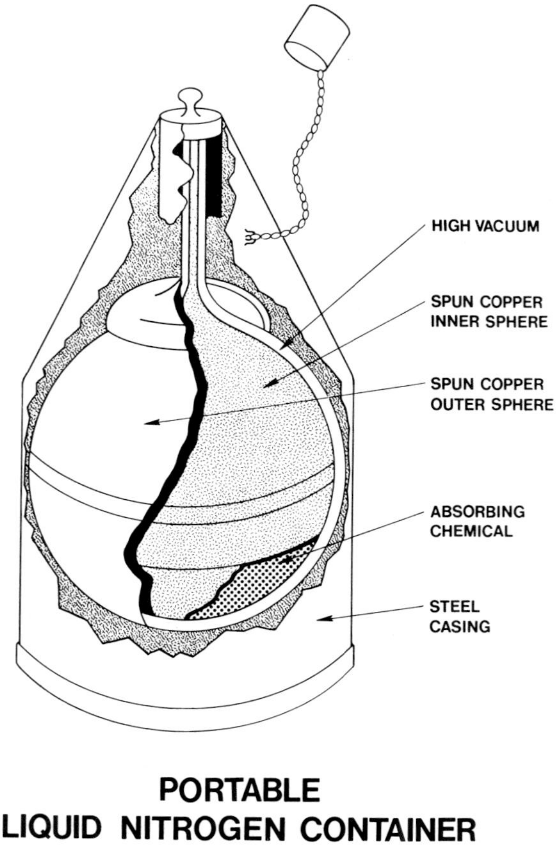

FIGURE 4. Schematic view of a typical portable Dewar. (Courtesy of Liquid Carbonic Corporation, Chicago.)

such as the one shown in Figure 8 becomes exceptionally practical and convenient. The mobile vessel contains all of the features required to maintain and dispense a large volume of cryogenic liquid or gas. Figure 9 illustrates the schematic of the vessel and provides the description of the components.

B. On-Site Storage Vessels

The final use of the cryogen usually will determine the size and pressure rating of the cryogenic vessel. Obviously, when it is desired to have long storage time with small losses, then a vessel with a relatively high design pressure of approximately 17.6 kg/cm^2 (250 psig [17 at]) would be installed. The higher working pressure of the vessel allows more vapor, which is produced from the constantly boiling cryogenic liquid, to collect and be held in the vapor space of the vessel. However, as the amount of vapor increases, the vessel pressure correspondently increases until it reaches the maximum working pressure of the vessel, after which the vapor is released to the atmosphere.

In comparison, a vessel with comparable capacity and design for low pressures of approximately 4.6 kg/cm^2 (65 psig [4.42 at]) would not be able to hold as much vapor as the higher pressure vessel. Therefore, the low-pressure vessel would be constructed from material with a thinner wall thickness and would contain less mass, which results in a lower heat leak rate on the vessel. Low-pressure vessels are suitable for installations in which the cryogenic liquid is to be used as a liquid for low-temperature applications such as cryogenic grinding, food freezing, and test chambers.

FIGURE 5. Actual external view of a portable Dewar. (Courtesy of Liquid Carbonic Corporation, Chicago.)

C. Vessel Operation

During normal filling operations, there is a rise in the vapor pressure of the vessel. This increase in pressure is caused by the rise in the liquid level, which decreases the vapor space available thereby compressing the same amount of vapor into a smaller volume. It is important to control or limit the amount of increase in pressure during the filling operation. To prevent the loss of vapor to the atmosphere, a recommended procedure is to "top fill" the vessel. This technique allows the incoming cryogen, which is at a slightly colder temperature than the liquid in the vessel, to condense some

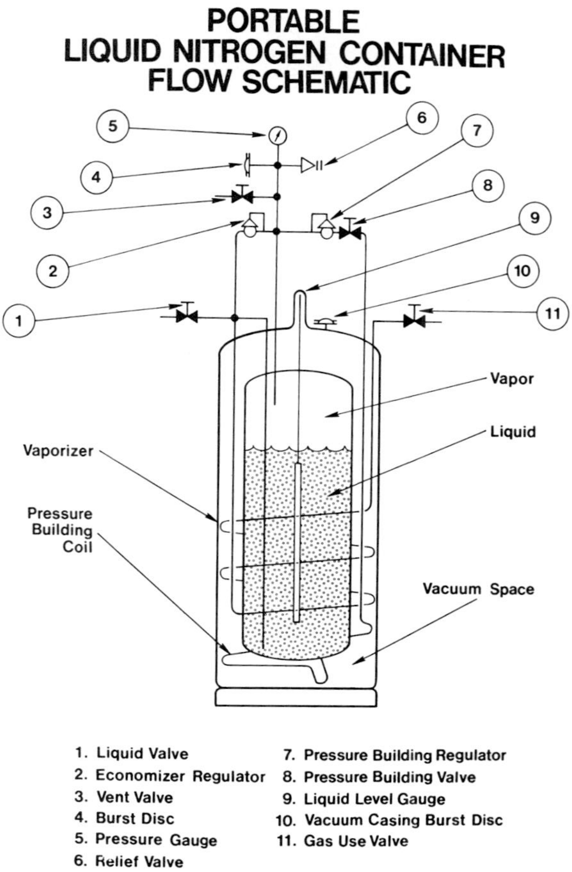

FIGURE 6. Schematic view of a pressure type portable Dewar (Courtesy of Liquid Carbonic Corporation, Chicago.)

of the vapor as it is sprayed into the vapor space. The use of a top-fill spray line effectively reduces the vapor pressure and eliminates any loss to the atmosphere by venting.

The content gauge should be used for estimating purposes only. Most content gauges used on cryogenic vessels operate on a pressure differential basis and require a stabilization period prior to reading due to the boiling action of the cryogen. This boiling action can cause erratic visual readings on the gauge and, therefore, would not indicate a true or accurate measurement of the contents of the vessel.

Each cryogenic vessel is equipped with some basic components which are necessary to maintain the vessel in a safe operating condition and allow for the withdrawal of either liquid or vapor. Once again, reference is made to Figure 3, which schematically shows the components of a typical cryogenic vessel.

The typical vessel can be used to dispense a liquid cryogen or gaseous vapors. If liquid is to be used, then a well-insulated or vacuum line is used to transfer the cryogen from the vessel to the use point.

In liquid cryogenic applications, the pressure in the storage vessel should be maintained as low as possible and be sufficient to overcome the pressure drop in the line and maintain adequate flow rates. The temperature of the liquid cryogen is directly related to the pressure of the storage vessel. As the pressure in the vessel increases, the

FIGURE 7. Typical 200 liter portable Dewar. (Courtesy of Liquid Carbonic Corporation, Chicago.)

corresponding temperature of the liquid also increases. For example, liquid nitrogen stored in a vessel at 0 psig pressure would have a temperature of −195.8°C (−320.4°F). However, if the same liquid nitrogen is stored in a vessel at 75 psig, it would have a temperature of approximately −173.33°C (−280°F). It is, therefore, extremely important to maintain the lowest possible vessel pressure for the application so that the maximum amount of available refrigeration from the liquid nitrogen can be used. The change in available refrigeration due to pressure changes is shown in Table 1 and Figure 10.

It may be necessary to use the pressure-building coil to maintain an adequate vessel

FIGURE 8. Typical mobile portable cryogenic vessel. (Courtesy of Liquid Carbonic Corporation, Chicago.)

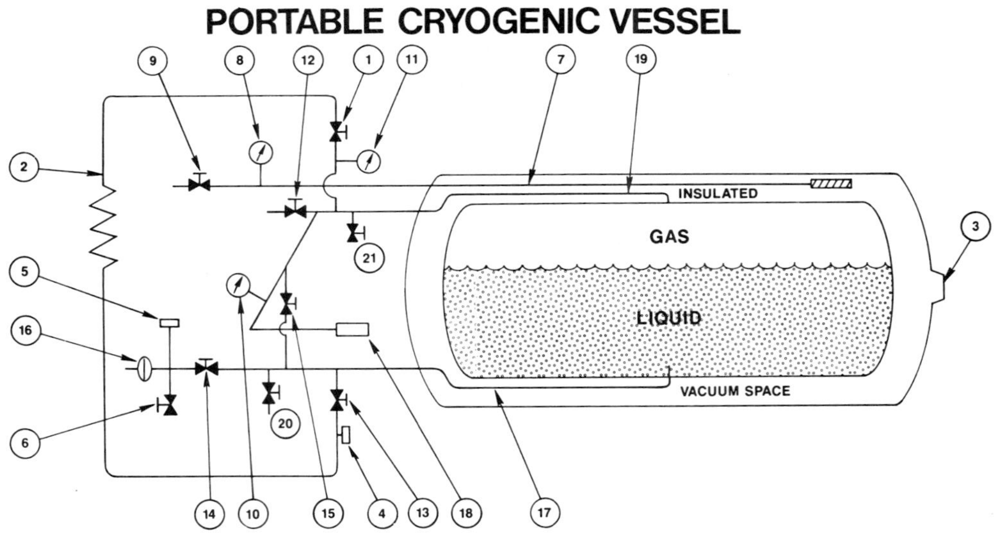

ITEM NO.	DESCRIPTION	ITEM NO.	DESCRIPTION
1.	Gas Pressure Build-Up Valve	12.	Vent Valve
2.	Pressure Build-Up Coil	13.	Pressure Build-Up Valve
3.	Rupture Disc Outer Tank	14.	Fill-Drain Valve
4.	Safety Disc	15.	Capacity Gauge Valve
5.	Safety Disc	16.	Filter
6.	Blow Down Valve	17.	Fill-Drain Line
7.	Vacuum Line	18.	Dampener Tank
8.	Thermocouple / Vacuum Gauge	19.	Vent Line
9.	Vacuum Valve	20.	Liquid Use Valve
10.	Capacity Gauge	21.	Gas Use Valve
11.	Pressure Gauge		

FIGURE 9. Schematic of a mobile cryogenic vessel. (Courtesy of Liquid Carbonic Corporation, Chicago.)

Table 1
TEMPERATURE/PRESSURE VS. LATENT HEAT OF FUSION NITROGEN

Temperature °F	Temperature °C	Heat of vaporization (Btu/lb)	Pressure (psia)	Heat of vaporization (Btu/lb)
−320	−195.7	86	14.7	86
−310	−190.0	81	20.0	84
−300	−184.4	76	30.0	80
−290	−178.8	72	40.0	78
−280	−173.3	65	60.0	75
−270	−167.7	62	80.0	72
−260	−162.2	56	100.0	66
−250	−156.6	48	200.0	58
−240	−151.1	32	300.0	48
−232.8	−147.1	0	400.0	34
			492.2	0

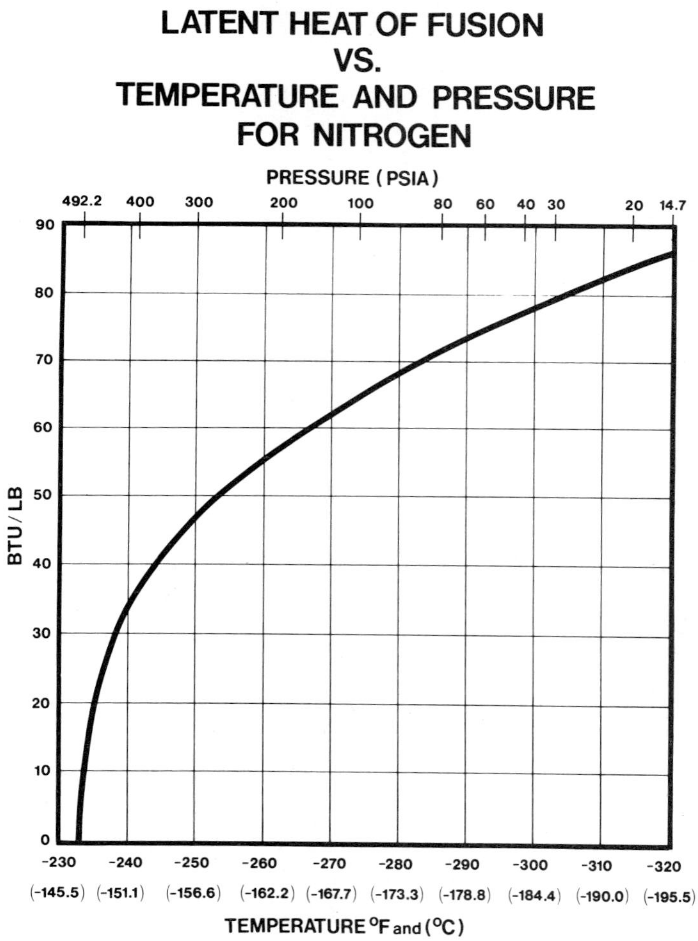

FIGURE 10. Latent heat of fusion vs. temperature and pressure for nitrogen. (Courtesy of Liquid Carbonic Corporation, Chicago.)

pressure during the time when liquid is being withdrawn from the vessel. For example, when the liquid cryogen is withdrawn at a greater rate, the liquid level in the vessel decreases, thereby creating a larger vapor space which in turn allows the original vapor pressure to decrease. To compensate for this condition, a portion of the liquid cryogen from the vessel is allowed to flow through the pressure building coil, which is essentially a fin tube heat exchanger. Heat from the atmosphere is used to transform the liquid cryogen into a cold gas, which is allowed to return to the vapor space in the vessel. With the use of the vapor pressure regulating device, the flow and transformation of the cryogen to a vapor is automatic and maintains a constant preset pressure in the vessel.

The typical cryogenic storage vessel can be used for a dual purpose; that is, it can dispense a cryogen to be used as a liquid or gaseous vapor. As previously described, liquid can be withdrawn by the use of insulated transfer lines. When there is a need for a gaseous vapor, a limited amount of vapor can be withdrawn directly from the head space of the vessel through a simple economizer line. However, when a substantially large volume of gaseous vapor is required, it becomes necessary to withdraw liquid and transform it to a vapor with the use of an external vaporizer. The vaporizer is basically a heat exchanger in which heat is transferred to the cryogenic liquid allowing the liquid to change to a warm vapor. The volumetric conversion from a liquid to a gas is different for each cryogen, as shown in Table 2.

The combined use of the vaporizer and the economizer system allows for the automatic withdrawal of liquid and vapor from the vessel. Vapors are withdrawn when the vessel pressure is above the preset economizer regulator setting. During this operation, liquid does not flow into the vaporizer. However, when the pressure in the vessel is reduced to the preset regulator setting, the flow of vapor from the head space stops and the liquid cryogen is allowed to flow into and through the vaporizer, thereby producing extremely large volumes of gaseous vapor for ultimate use. The adaptation and use of an economizer system will generally prevent any venting losses from occurring to a vessel.

D. Use of Load Cells

During recent years, the adaptation of load cells to cryogenic vessels has become an acceptable method for determining the contents of the vessel. Load cells can be used to accurately weigh both liquid and gaseous vapor within the vessel. The system will provide accurate scale readings and will not be affected by a difference in pressure or turbulence caused by the boiling cryogenic liquid. Load cells can be used as an addition to or a replacement for the commonly used liquid level gauge generally installed on cryogenic vessels.

A typical load cell system for a vessel with three legs is shown in Figure 11. The weighing system consists of individual ''cell'', connecting cable, and a readout device. When a downward force, which can be considered as the unknown weight, is applied to the cell device, a small amount of deflection occurs thereby producing a change in the electrical output signal. This electrical circuit change is then used to produce a visible change in numerical weight readings on the display panel. Any change in the amount of deflection of the cell device caused by an increase or decrease in the weight of the vessel will be correspondently recorded in the display panel.

The accuracy of the system is greatly improved when load cells are used under each supporting leg of a vessel. A typical industrial cryogenic installation is shown in Figure 12 in which three cells are used together with the connecting electrical cable and visible readout panel. Accurate weight readings of the cryogenic liquid in the vessel are instantaneously available even when liquid is being withdrawn from the vessel.

Table 2
CRYOGEN VOLUMETRIC CHANGE

Liquid to vapor @ 1 Atm and 21.1°C (70°F)

°F	°C	Cryogen	Volume change
−320.4	−195.8	Nitrogen	696.5
−297.4	−183.0	Oxygen	860.6
−320.5	−185.8	Argon	841.2
−423.2	−252.9	Hydrogen	848.1
−454.0	−268.9	Helium	754.2

E. Evaporation/Heat Leak/Vacuum

Cryogenic liquids are considered boiling liquids at temperatures substantially colder than the surrounding atmosphere. As the liquid boils, it produces gaseous vapor. This physical change of state from a liquid to a gas will continue because of the temperature difference between the cryogenic liquid and the surrounding atmosphere. Heat which is transferred from the warmer atmosphere to the cold liquid causes the liquid to boil and vaporize. There is no practical way to prevent completely the transfer of heat through a storage container to the cryogenic fluid. It therefore becomes necessary to design the storage vessel so that only the minimum amount of atmospheric heat is transferred to the boiling liquid.

Manufacturers who fabricate spherical, vertical, or horizontal vessels, as shown in Figure 13, strive to have the least number of directly connected structural members between the inner and outer shell. This design technique, together with the insulation material used and the surrounding vacuum, determines the "heat leak" of the cryogenic vessel. The amount of evaporation that will occur in any period of time is determined by the heat per unit of time that passes through the insulation, supports, vacuum, and piping, and is absorbed by the cryogenic fluid.

Manufacturers of cryogenic vessels use the most modern welding techniques and equipment available to fabricate the vessels with high-quality seams to ensure the vacuum integrity of the vessel. The most minute imperfection in the weld seam will result in a continuous degradation of the vacuum. Figure 14 illustrates one method which is used to produce the X-ray-quality weld required to maintain a vacuum condition. A typical 42,770-ℓ (11,300-gal) cryogenic vessel is shown in Figure 15.

Heat leak and evaporation rates can be different for vessels of comparable capacity, when made by different manufacturers. However, manufacturers and suppliers of cryogenic vessels can usually provide approximate evaporation rates for their vessel. The evaporation rates are given as a percentage of the total volume during each 24-hr period. Usually, the smaller the size of the vessel, the higher the evaporation rate as compared to larger vessels. Table 3 illustrates the approximate nitrogen evaporation rates for various size cryogenic containers and larger industrial-type vessels.

It is important to understand that the vapors generated by the normal heat leak of a vessel need not be lost or vented to the atmosphere. These generated vapors are normally allowed to build up in the head of the vessel, continuously increasing the pressure until it reaches the setting of the relief device. Normally, the pressurized vapors are withdrawn from the vessel and used for in-plant processing thereby eliminating or minimizing any loss. By utilizing this technique, many vessels seldom vent any vapors to the atmosphere because there is a continuous withdrawal of vapors from the vessel, which is equal to or greater than the evaporation rate of the vessel.

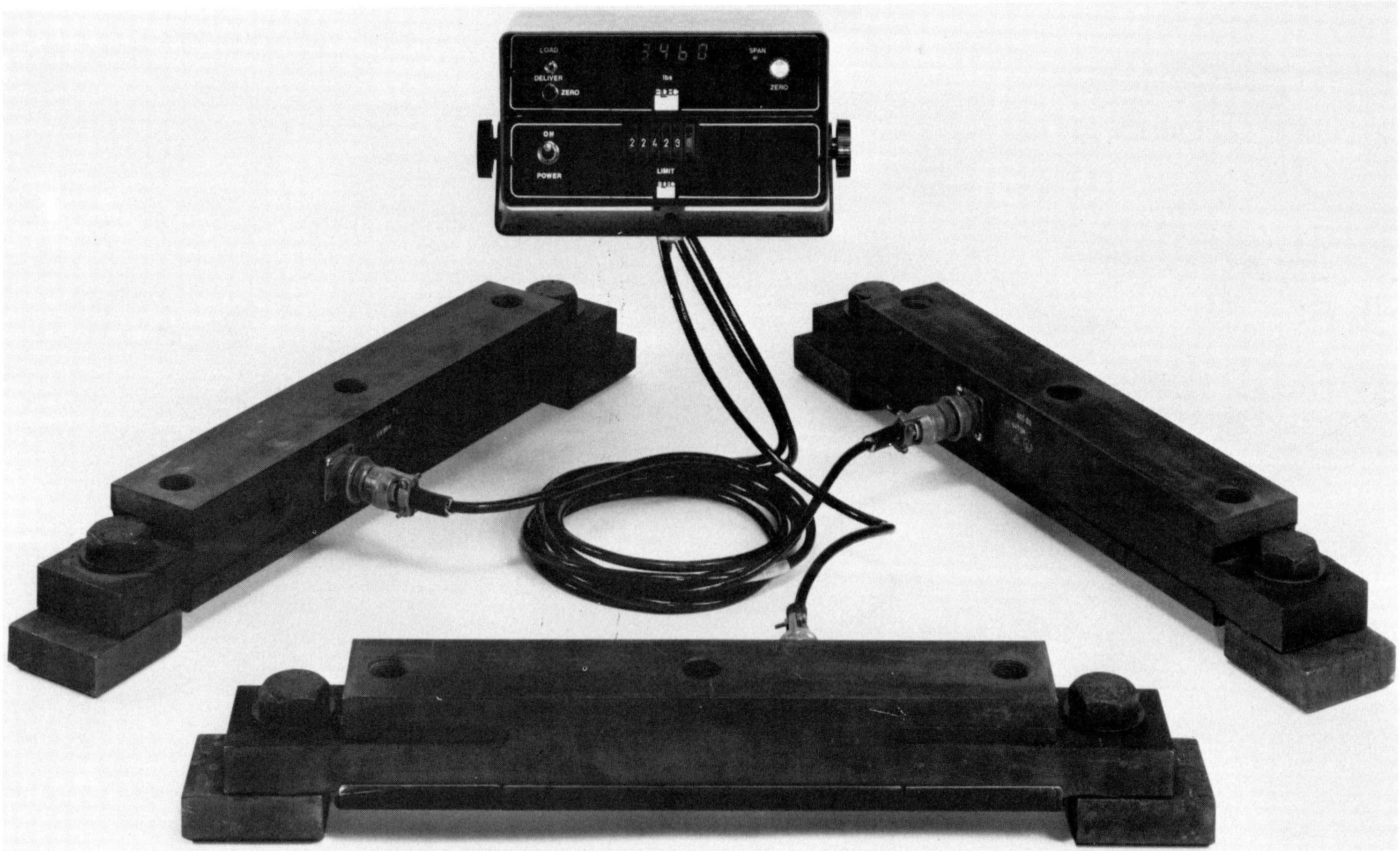

Figure 11. Typical load cell system. (Courtesy of Eldec Corp. Industrial Division, Lynnwood, Wash.)

FIGURE 12. Typical load cell installation on a cryogenic vessel. (Courtesy of Liquid Carbonic Corporation, Chicago.)

FIGURE 13. Various types of cryogenic vessels being fabricated. (Courtesy of Process Engineering, Inc., Plaistow, N.H.)

FIGURE 14. Unique seam welding of a cryogenic vessel. (Courtesy of Process Engineering, Inc., Plaistow, N.H.)

Large-capacity cryogenic vessels are fabricated with a vacuum space between the inner vessel and outer protective shell. The evaporation (boil-off) rate is directly related to the degree of vacuum in the annular space. For example, as the vacuum level becomes lower, the amount of evaporation correspondingly becomes smaller. Figure 16 shows the anticipated evaporation rate for a typical 37,850-ℓ (10,000-gal) low-pressure vessel when compared to the vacuum condition of the vessel.

It can be seen from Figure 16 that for each reduction of 40 μm, the evaporation loss is decreased by 45.3 kg (100 lb) of nitrogen per 24 hr. The chart should be used only as a guide as the evaporation (loss) values are approximate and will change based on vessel design and type of insulation material used.

The vacuum condition that is maintained in conjunction with the insulation is considered to be a relatively high degree of vacuum. The condition of the vacuum is measured in "micron" units. A micron is a unit of vacuum equal to 1/1000 of a millimeter of mercury pressure. At atmospheric pressure a vacuum gauge will read 760 mm, and at absolute vacuum it would read zero. A vacuum pump is used to evacuate the vessel's annular space, and as the air is removed, the vacuum gauge readings begin to descend from 760 mm to 1 mm. The last 1 mm of pressure is graduated into 1000 "micron" readings. As the vacuum pump continues to operate, the micron readings decrease until the limitation of the pump is reached.

Figure 17 shows one type of a portable vacuum gauge used to determine the vacuum condition of typical cryogenic storage vessels. Normally, a portable vacuum gauge as shown is used in preference to individual vessel-mounted gauges. Table 4 shows the conversion values used for various units when determining vacuum conditions. The simple and quick way of determining the vacuum condition of a cryogenic vessel is shown in Figure 18.

FIGURE 15. Completed 42,770-ℓ (11,300 gal) vessel. (Courtesy of Process Engineering, Inc., Plaistow, N.H.)

Table 3
APPROXIMATE NITROGEN EVAPORATION RATES FOR COMMERCIALLY AVAILABLE EQUIPMENT

Portable container size (Gal)	(*l*)	N_2 gas equivalent	%/day	CF/day
0.5	2	49 CF	100.0	49.00
1.3	5	123 CF	7.0	8.61
2.6	10	246 CF	4.0	9.84
29.0	110	2706 CF	2.0	54.12
42.2	160	3936 CF	2.4	94.46
Storage vessel size				
(Gal)	**(*l*)**			
500	1,892	46,555 CF	0.5	232.7
900	3,406	83,799 CF	0.4	335.1
1,500	5,677	139,665 CF	0.4	558.6
3,000	11,355	279,330 CF	0.35	977.6
6,000	22,710	558,660 CF	0.3	1675.9
11,000	41,635	1,024,210 CF	0.25	2560.5

Note: CF, equivalent conversion of LN_2 to vapor.

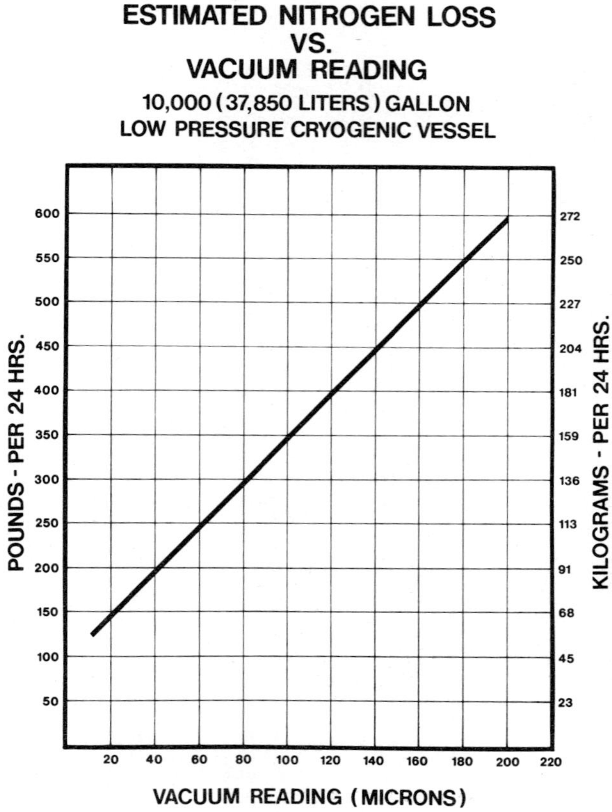

FIGURE 16. Nitrogen loss vs. vacuum reading 37,850 liter vessel. (Courtesy of Liquid Carbonic Corporation, Chicago.)

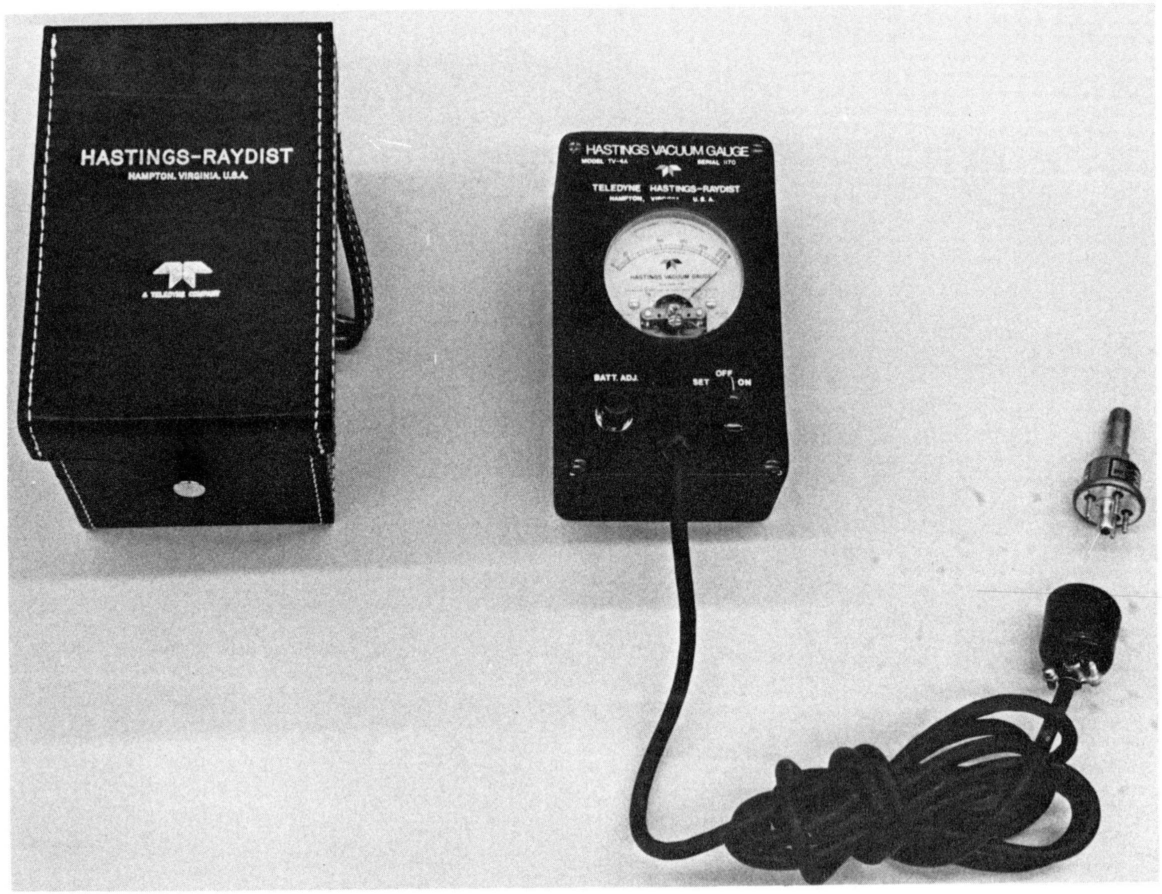

FIGURE 17. Typical portable vacuum gage and element. (Courtesy of Teledyne Hastings-Raydist, Hampton, Va.)

Table 4
VACUUM CONVERSION TABLE

Units		Pumping system: Single stage mechanical pump			2-stage mechanical pump (MICROtorr 1)			MICROtorr 11C	
Torr	760	25.4	1	10^{-1}	10^{-2}	10^{-3}	10^{-4}	10^{-5}	10^{-6}
Microns	760,000	25,400	1000	100	10	1	.1	.01	.001
Inches Hg	29.9	1	.04	.004	.0004	—	—	—	—
Atmospheres	1	.034	.0013	.0001	—	—	—	—	—

Typical industrial-designed cryogenic storage vessels normally have a vacuum condition of 100 to 200 μm during a warm condition. When the vessel is initially filled with a cryogenic liquid, the temperature within the inner vessel and the annular vacuum space begins to decrease. During this same period of cooldown, the vacuum reading also decreases. After the temperatures have stabilized, it is fairly common to have the vacuum reading reduced to 10 to 20 μm in a cold condition. The extremely low vacuum reading will normally remain relatively constant for a considerable period of time providing the integrity of the vessel is maintained.

The heat leak value of a cryogenic vessel is also directly related to the pressure rating of the vessel. Obviously, when a vessel is fabricated for higher working pressures, the thickness of the metal used will be greater, the total mass or weight will be more, and therefore the expected heat leak will be higher than a lower working pressure vessel with approximately the same volume. As an example, compare a nominal 34,822-ℓ (9200-gal) vessel which holds approximately 34,065 ℓ (9000 gal) of liquid and a 10% vapor space, with a working pressure of 3.5 kg/cm^2 (50 psig [3.4 at]), to a similar vessel with a working pressure of 15.8 kg/cm^2 (225 psig [15.3 at]) — each with a 50 μm vacuum reading. The lower pressure-rated vessel would generate approximately 1% less of the expected evaporation rate per 24 hr of the higher designed pressure-rated vessel.

The larger amount of evaporation is partially attributed to the increase in metal thickness of the vessel which is required for the higher working pressure of the vessel. In addition, it may be necessary to increase the size of the supporting members, thus allowing more heat to be transferred to the cryogenic fluid.

F. Cryogenic Piping

Cryogenic transfer lines as described in this section are based on current acceptable use in industrial applications. The types of transfer lines can easily be divided into three categories:

1. Uninsulated metal lines
2. Insulated lines with cellular material covering
3. Vacuum-insulated lines

The use of an uninsulated metal line for "temporary" installations or short flexible connections is sometimes the simplest method to use as a transfer line. Care should always be exercised when exposed piping is used on a temporary basis. Obviously, frost will form on the surface of the transfer line due to the extreme temperature difference between the cryogenic liquid and the atmosphere. Uninsulated lines are commonly used when transferring liquid oxygen or nitrogen between mobile road transport trailers and on-site storage vessels. Transfer lines of this type are usually sized to handle

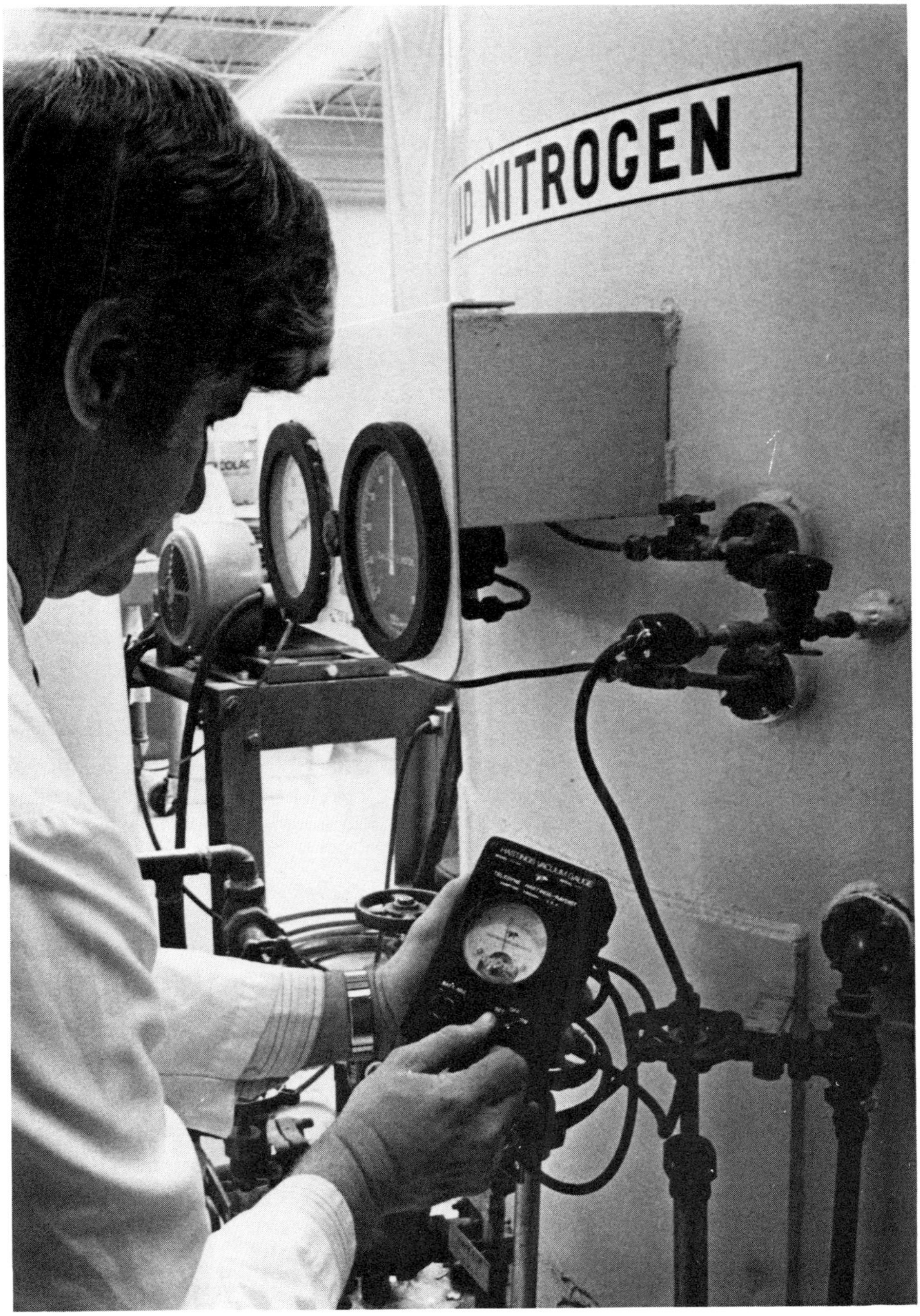

FIGURE 18. Vacuum reading being determined on a cryogenic vessel. (Courtesy of Liquid Carbonic Corporation, Chicago.)

large flows for short periods of time. End fittings are different in design and configuration for each type of cryogenic liquid to prevent the accidental mixing of different gases.

It can be assumed that under certain conditions air can condense on the outside surface of an uninsulated transfer line which has liquid nitrogen flowing through it. Although the condition may exist for only a brief period of time, uninsulated lines may prove hazardous under certain circumstances. Hence, uninsulated lines should be used with caution — and only temporarily — for short periods of time.

Conventional, cellular-type insulation such as polyurethane has become widely acceptable as a relatively low-cost cryogenic insulating material. Preinsulated pipe is commercially available and can be obtained with the specific radial thickness of insulation desired. This type of piping usually will consist of an internal copper pipe which is surrounded by two, three, four or more inches of polyurethane insulation and then enclosed within a polyvinylchloride (PVC) cover. A typical section of commercially available preinsulated pipe is shown in Figure 19. The PVC cover acts as the vapor barrier which prevents water vapor or ice formation from degrading the insulation. Adaptor pieces such as elbows, tees, valve bodies, straight line connectors, etc. are also commercially available. The cooldown loss associated with preinsulated piping is more than it is with uninsulated lines because more mass such as the insulation has to be cooled down. However, the steady state heat leak will be substantially lower. An approximate heat leak comparison is shown in Table 5. As in all cryogenic transfer lines, allowance must be made for the contraction and expansion of the pipe due to extreme temperature changes.

Vacuum-type transfer lines are usually considered when there is substantial distance between the storage vessel and the use point, and when large and almost continuous flow is required. Under these circumstances, the reduced amount of heat leak of a vacuum transfer line may compensate for the additional cost as compared to preinsulated piping. Usually, vacuum piping would be considered for distances in excess of approximately 120 m (400 ft). However, careful cost comparison should be made before any type of transfer line is selected.

Vacuum transfer lines are available in two types: rigid and flexible. General comparison appears to show that the rigid type is substantially higher in cost due to the method of fabrication. As the name implies, rigid vacuum piping is made in specific lengths of stainless steel material. Bellows are used for the expansion and contraction of the pipe with all of the connecting areas being joined using an inert gas-tungsten-welded steam, and later tested for possible leaks. The rigid vacuum transfer line can have a static vacuum condition with connections in each section for future vacuum pumping if required. Usually rigid vacuum transfer lines do not provide much flexibility for future changes without expensive modifications.

Flexible vacuum transfer lines are available in various sizes and lengths. This type of transfer line provides a relatively large degree of flexibility at a substantially lower cost. The line is normally supplied in one continuous length when applicable, constructed of concentric inner and outer corrugated copper tubes with a polyvinylchloride sheath.

The design of the vacuum line is such to allow for bending, contraction, and expansion. A typical section is shown in Figure 20. Manufacturers of flexible vacuum lines suggest that a vacuum pump be included as part of each installation to maintain the vacuum at the desired level.

FIGURE 19. Commercially available preinsulated cryogenic pipe. (Courtesy of Liquid Carbonic Corporation, Chicago.)

Table 5
APPROXIMATE HEAT LEAK COMPARISON FOR CRYOGENIC PIPING

Approximate Heat Loss for Liquid Nitrogen — Cryogenic Piping — 1-In. Nominal O.D. Inner Pipe

Type of insulation	Btu/hr/ft	Lb/hr
1-in. uninsulated pipe	190	2.2
4-in. radial cellular block insulation	39	0.45
2-in. radial polyurethane insulation: PVC sheath	11	0.12
3-in. radial polyurethane insulation: PVC sheath	9	0.10
4-in. radial, rigid vacuum powder insulation including end connections	6	0.07
4-in. radial vacuum insulated with laminar shield	3	0.03
2-3/8 in. radial flexible vacuum line at 1×10^{-5} torr vacuum	1.8	0.02

Note: All values are approximate due to variation in manufacture, density of insulation, etc.

G. Cryogenic Transfer Line Design

There are some basic considerations to be followed in the design and installation of every cryogenic transfer line. Compliance to the basic principles will provide a safe, economical, and functionable system. Some of the principles are as follows and are illustrated in Figure 21.

1. Select the type of insulation suited for the specific installation.
2. Use a pipe size adequate to deliver the amount of cryogen required with the lowest acceptable pressure drop.
3. Use extended stem valves designed for cryogenic service.
4. Use only acceptable material such as austenitic stainless steel, copper, bronze, aluminum.
5. Provide and use adequate expansion joints to allow for contraction.
6. Use approved safety pressure relief valves in every section between any two block valves or other devices which can trap liquid between them.
7. Use risers between the relief valves and the liquid line to provide some heat transfer to the relief device, thereby preventing a freeze-up.
8. Never use plain carbon steel material because of failure due to low-temperature embrittlement.
9. Avoid gas traps and slope the line upward from the vessel.
10. Use a phase separator when there is an instantaneous need for liquid at a use-point.

H. Automatic Cryogenic Controls

There are numerous electrical and pressure controls available to assist in the designing of automatic cryogenic dispensing systems for low temperature changers or liquid nitrogen batch cooling systems. The adaptation of basic controls for various needs are shown in Figure 21.

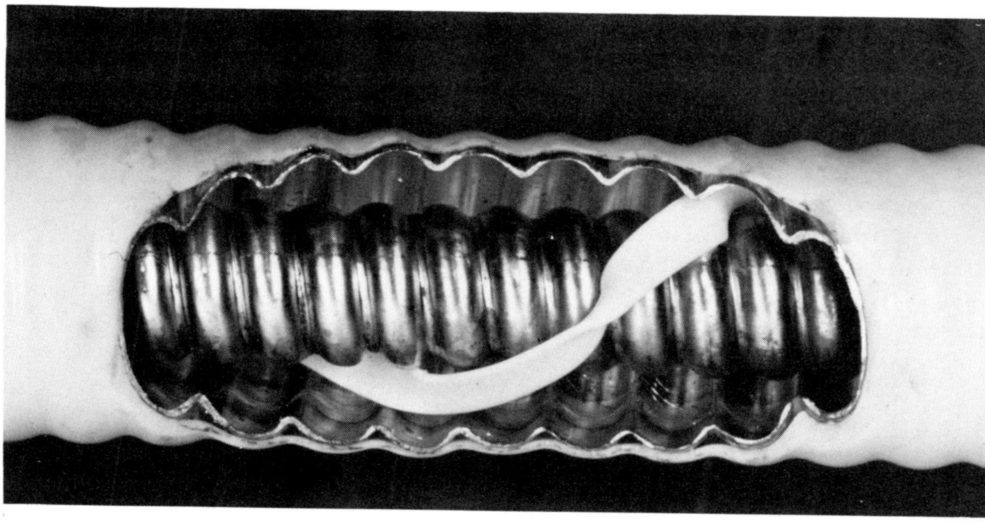

FIGURE 20. Flexible vacuum pipe. (Courtesy of Vacuum Barrier Corporation, Woburn, Mass.)

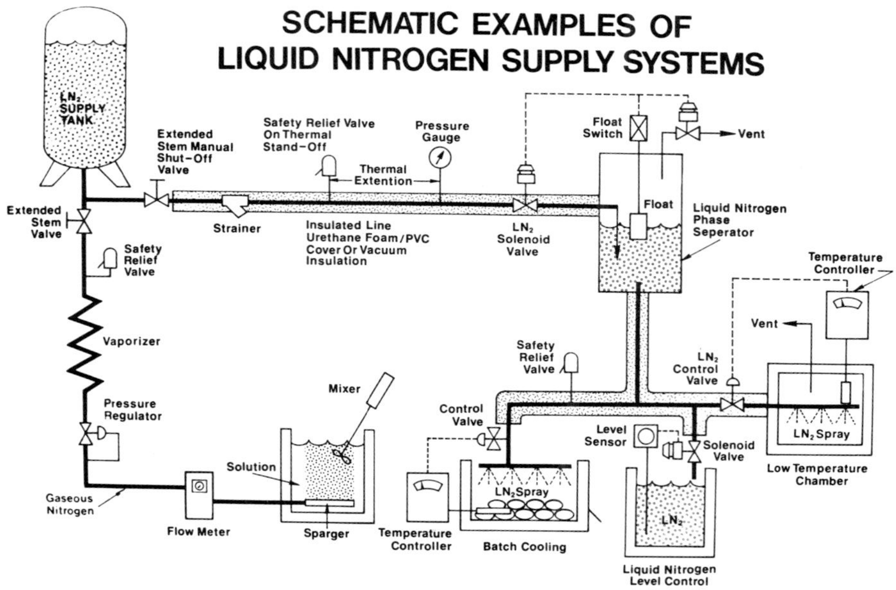

FIGURE 21. Schematic examples of liquid nitrogen supply systems. (Courtesy of Liquid Carbonic Corporation, Chicago.)

III. CRYOGENIC SAFETY PRECAUTIONS

Cryogenic liquids such as liquid nitrogen can be extremely useful in developing new industrial processes and research studies. However, due to the extremely low temperature of the cryogen, it can become potentially hazardous if simple safety precautions are not taken. The following is a summary of the considerations which should be addressed when dealing with cryogenic liquids:

Personal safety — Extremely cold temperatures can quickly freeze (cryogenic burn) unprotected parts of the body, such as eyes, hands, face, or other exposed skin areas. Precautions should be taken to prevent a cryogen such as liquid nitrogen from coming into contact with the body. Use protective eye shields, proper gloves, shoes, and apron. Do not come into direct contact with material such as piping that has been cooled down with liquid nitrogen.

Storage equipment — Cryogenic storage equipment should be installed in a manner to comply with National Fire Protection Association codes and other local building codes. If cryogenic storage vessels are installed in confined or closed areas, proper vent piping should be used in conjunction with safety relief valves which exhaust the venting nitrogen to the atmosphere.

Acceptable materials — Many materials such as plastic, cloth, and rubber become weak and reach their embrittlement point when subjected to low temperatures. Materials such as plastic, glass, concrete, etc. can easily crack or shatter due to thermal shock when subjected to liquid nitorgen. Ordinary carbon steel, most alloy steels, and the 400 series stainless steels should not be used for cryogenic service as they easily fracture at low temperatures. Copper, brass, bronze, monel, aluminum, and the 300 series stainless steels are satisfactory for cryogenic systems as they remain relatively ductile at cryogenic temperatures.

Piping — Conventional screw type pipe joints can be used with Teflon® tape joint sealant. However, it should be remembered that as the piping becomes cold, the material shrinks. This condition can cause leaks at the joints. Therefore, care should be taken to make up a tight connection at first. The use of soft solder is acceptable for slip-fit joints when used for short periods of time and low pressure operations. Soft solder joints will fail in time due to cracks produced from temperature cycling. Silver solder with at least 44% silver or more is preferred and should be used when making permanent slipfit joints in copper tubing. Welding of acceptable stainless steel pipes is very satisfactory. Always provide for adequate expansion and contraction of the piping.

Open containers — Care should be exercised when using large open top Dewars as oxygen from the atmosphere can condense in the container. The resultant liquid may then be capable of supporting combustion.

Pressure vs. vacuum — Extreme care should be used to prevent the trapping of liquid nitrogen between two valves or any other devices. Liquid nitrogen expands to almost 700 times that of its liquid volume. This tremendous change from a liquid to a gaseous vapor state can produce extremely high pressures and cause explosions. To prevent this type of accident from occurring, proper relief devices should be installed in the piping wherever liquid nitrogen can possibly be trapped. It is recommended that a stainless steel extension tube be vertically installed between the transfer pipe and the relief valve. This type of installation allows gas to accumulate just below the relief valve and be somewhat near ambient temperature, thereby allowing the device to function when required. It is also possible to cause an implosion on equipment which is cooled by liquid nitrogen. Due to the cold temperature of liquid nitrogen, it is sometimes possible to create a sufficient vacuum which will cause the equipment to collapse. This possible condition can be avoided by the installation of a vacuum breaker on the equipment or enclosed vessel.

Oxygen concentration — There may exist the possibility that air can condense on the outside surface of lines which carry liquid nitrogen. This remote possibility may occur because the normal boiling point of nitrogen is −195.8°C (−320.4°F) and oxygen is −183.0°C (−297.4°F). If allowed to collect and form a pool, the oxygen-enriched liquid air will become further enriched in oxygen as the condensate boils away. If such a condition develops, the condensate should not be allowed to come into contact with combustible materials or allowed to become confined in any type of enclosure.

Asphyxiation — As a colorless, odorless, tasteless, and nontoxic gas, the use of heavy concentrations of nitrogen in confined areas can act as an asphyxiant. Special care should be used when handling nitrogen in confined areas. To prevent the depletion of oxygen, it may be necessary to force air into the confined area. An oxygen level of 19.5% or more should be maintained (21% is normal) if personnel are required to enter the area. Areas with heavy concentrations of nitrogen should be purged with air and tested to determine the oxygen level before personnel are allowed to enter the area.

Handling precautions — (1) Avoid skin contact with liquid nitrogen or related equipment. (2) Avoid contact between liquid nitrogen and materials not approved for such service. (3) Use approved type valves (extended stem) and safety devices. (4) Always wear loose fitting gloves and protective clothing. Clothing should be free of cuffs to avoid trapping or collecting liquid. (5) Wear adequate eye protection at all times. (6) Use suitable shielding when handling liquid in open containers to prevent the contact of liquid and atmospheric air. (7) Never trap liquid nitrogen in a container, closed system, or a pipe section without suitable pressure relief devices. (8) Provide adequate ventilation when handling liquid nitrogen in confined spaces. (9) Always allow for expansion and contraction in pipeline and container. (10) A thorough training of all personnel handling liquid nitrogen is of greatest importance.

REFERENCES

1. **McClintock, M.,** *Cryogenics,* Reinhold Publishing, New York, 1964.
2. **Scott, R. B.,** *Cryogenic Engineering,* D Van Nostrand, Princeton, N.J., 1959.
3. **Sittig, M.,** *Nitrogen In Industry,* D Van Nostrand, Princeton, N.J., 1965.
4. **Vance, R. W. and Duke, W. M.,** *Applied Cryogenic Engineering,* John Wiley & Sons, New York, 1962.
5. **Sittig, M. and Kidd, S.,** *Cryogenics Research and Applications,* D. Van Nostrand, Princeton, N.J., 1963.
6. **Liquid Carbonic Corporation,** Physical and Chemical Gas Data Chart, Form No. 6577, Liquid Carbonic Corp., Chicago, 1976.
7. **Liquid Carbonic Corporation,** Liquid Receivers, Tech. Sales Bull. No. IE-3003, Liquid Carbonic Corp., Chicago.
8. **Liquid Carbonic Corporation,** Portable Liquid Receivers, Tech. Sales Bull. Nos. IE-3004 and IE-3006, Liquid Carbonic Corp., Chicago.
9. **Liquid Carbonic Corporation,** Vacuum Gauges/Elements, Tubes, Tech. Sales Bull. No. IE-3023, Liquid Carbonic Corp., Chicago.
10. **Liquid Carbonic Corporation,** Cryogenic Pipes and Heat Leak Calculations, Tech. Sales Bull. Nos. IE-3016 and IE-3022, Liquid Carbonic Corp., Chicago, 1974.
11. **Liquid Carbonic Corporation,** Safe Handling of Compressed Gases, Form No. 6637, Liquid Carbonic Corp., Chicago, 1978.
12. **Union Carbide Corporation, Linde Division,** Nitrogen Brochure, Form No. F-4218, Linde Div., Union Carbide Corp., New York, 1978.
13. **Liquid Carbonic Corporation,** Nitrogen, Tech. Sales Bull. No. IG-2001-R, Liquid Carbonic Corp., Chicago.
14. **Vacuum Barrier Corporation,** Semiflex/Triax Bulletin No. AP-150C and No. SFA-100, Vacuum Barrier Corp., Woburn, Mass.

Chapter 8

THE USES OF CRYOGENICALLY RECYCLED RUBBER

J. Koutsky

TABLE OF CONTENTS

I. INTRODUCTION

Although cryogenic recycling is in its infancy as a recycling method, interest in this technique has been primarily spurred by the numerous uses of the products obtained by cryogenic pulverization. It must be emphasized that cryogenic milling is primarily a physical process, and the low temperature and nitrogen environment during milling creates products which have not been substantially oxidized or chemically changed. This allows recycling of even chemically sensitive organic materials or metals which easily oxidize. A prime example of cryogenic pulverization has been in the area of grinding spices which are especially delicate substances to oxidation.

This chapter will review some properties and uses of cryogenically milled products and will emphasize primarily recycled rubber.

II. CRYOGENICALLY RECYCLED RUBBER

Depending upon the configuration of the cryogenic mill, the amounts of liquid nitrogen used to process recycled rubber can vary. In experiments at the University of Wisconsin—Madison, about 0.45 to 0.63 kg (1 to 1.2 lb) of liquid nitrogen are used for milling 0.45 kg (1 lb) of tire rubber to particulates of 10 to 80 mesh sizes. The particles distribute about mean sizes of 40 to 50 mesh for a common milling operation. These particles can be molded or extruded by themselves or mixed with other materials. The particles usually have a density near that of water, are generally nonwetted by water, and absorb many organic liquids and even metal ions. These properties allow for many diverse uses of particulized rubber. Some of these uses are shown in Table 1.* These products are from rubber obtained from discarded tire carcasses, and the desired properties of each product will not be greatly affected by tire type.

A. Molded or Extruded Products

Cryogenically milled tire rubber generally gives particles as shown in Figure 1. The surfaces have both sheared and brittle fracture surfaces, with brittle fracture predominating. The particles are angular in nature and the surfaces have a high degree roughness. These particles can be compression-molded by themselves or mixed with processing aids such as softeners, fillers (waste wood fines from paper processes), and other polymeric materials.

Most tire particles begin to soften around 176° to 204°C (350° to 400°F). The strength of a molded sheet made from rubber particles alone, maximizes at about 233°C (450°F).[10] Figures 3 to 6 show typical data of molded sheets of cryogenically recycled rubber obtained from passenger tires pulverized to a distribution shown in Figure 2. Mold pressures must be maintained above 10^6 pascals (150 psi) to obtain the strongest sheets. The effect of particle size on the strength of the sheet is great. Generally, smaller particle size gives greater strengths. It is interesting to note, however, that sizes smaller than 80 mesh appeared to give weaker sheets since greater amounts of dirt and clay exist in these fractions.

Figure 7 and Table 2 show typical stress-strain properties of molded recycled rubber compared with "virgin" rubber products.[10] Simple compression molding gives much lower strengths and elongations for recycled rubber products when compared to natural rubber and SBR (styrene-butadiene rubber). To improve strength and maximum elongation the adhesion between particles must be increased, and with the addition of softeners, such as dodecylamine (DDA), elongations of the sheets can be increased to

* All tables and figures appear at the end of the chapter.

over 150% (Figure 8).[10] With the addition of curable rubberized cements to the rubber particles, an increase of tensile strength of 5 to 20 times can be realized with 5 to 20 weight percent cement addition. Molding times, pressures, and temperatures are optimized to achieve these strengths of the molded sheets.

Incorporation of these cements allows the molding of high strength-high elongation sheets which can be useful for floor coverings — and, in particular, large animal mats. These mats may be 5 to 7.5 cm (2 to 3 in.) thick and over 4.6 to 6.5 m^2 (50 to 70 ft^2). The mold times are generally quite long (1 to 2 hr) to achieve uniformity throughout the thickness of the mat.[2]

Incorporation of solid fillers increases the modulus of the molded rubber but generally decreases the maximum elongation. Typical data are shown in Figure 9 for wood fine filler which came from washed residues obtained from a Kraft® paper plant. The modulus and shore hardness of the molded rubber sheet are considerably increased by addition of wood fines.

The addition of other polymeric materials which are fusible (i.e., thermoplastics) considerably increases the strength of the molded sheet material. It is essential that the softening point of the thermoplastic and rubber are both exceeded during the molding operation. Figures 10 and 11 show the influence of polyethylene and polypropylene additions to the rubber particles molded at 3.1×10^6 pascals (450 psi), 233°C (450°F) for 5 min for a 0.178-cm (1.8-mm) sheet thickness.[10] With these additions, considerable improvement of tensile strength allows these compression molded products to find such applications as flooring sheets and lawn mower blades.[4]

Improved bonding or compatability between the rubber particles and thermoplastics increases strength and elongation of the molded sheet. For example, the rubber particles can be surface-modified by chemical grafting of adhesive materials such as polyvinyl acetate and incorporate the modified rubber particles into a poly vinyl-acetate-ethylene co-polymer to give high strength-high elongation molded sheet materials.[9]

Extrusion of recycled rubber can be facilitated by the addition of processing aids such as oils which absorb and soften the rubber. The product can vary in porosity depending upon processing pressure, speed, temperature, and additives. Some volatile materials can be added purposefully to create bubbles in the product, such as flexible rubber foams for padding and insulation. An interesting application of recycled rubber extrusion is the production of porous hose ("leaky hose") used for irrigation purposes.[1] The original additives incorporated into the tire rubber inhibit the chemical degradation of the hose, so the outdoor environmental exposures of the hose are of minimal consequence. Very high strengths are not needed for such a hose since water pressures would be modest.

Other applications of extruded cryogenically recycled rubber are in extrusion blends of rubber and polyethylene or polypropylene. Very durable fence posts have been extruded which can be inserted directly into the ground without worry of termites and fungi.[1] These posts can be nailed to other cross-pieces, as in ordinary fencing material, and no paint coating is needed to protect the post from weathering.

Coatings, paints, and sealants can be made from the recycled rubber particles. The particles are dispersed and dissolved in a suitable hydrocarbon, such as toluene or other asphaltic blends.[2] To achieve best results the rubber particle sizes must be 50 mesh or smaller to achieve product uniformity.

Rubberized asphalt pavements and seal coats offer an important use for recycled rubber particles. As has been described by McDonald and Olsen,[5] the recycled rubber of 16 to 25 mesh size is added to penetration grade asphalts in amounts up to 40 weight percent rubber. Usual amounts are about 25%. The mixture is blended at 149° to 232°C (300° to 450°F) to a thickened mixture. (At room temperature the material is an excellent tough binder). The hot mixture is applied to the original cracked pavement at a rate of 1.8 to 3.8 × 10^{-3} m^3/m^2 (0.4 to 1 gal/yd^2). After the application of the rubber-asphalt binder, a cover stone 0.95 cm (3/8 in.) nominal size is applied at about 21 kg/m^2 (38 lb/yd^2) and subsequently embedded into the binder by rolling. This rubber-asphalt binder seal prevents cracking, acts as a moisture barrier, inhibits bleeding of the asphalt in hot weather, and gives less brittleness at cold temperatures.

Cryogenically recycled rubber particles absorb many organics to a significant extent.[12] The amount of absorption is simply defined as the mass of absorbed organic per mass of original rubber particle. Generally this value is greater than two and depends upon the size of rubber particles, temperature, and the specific nature of the organic. Polar organic materials generally are not as greatly absorbed as nonpolar organics. Tables 3 and 4 summarize the absorption values for a variety of organic compounds. For particles larger than 30 to 40 mesh, absorption values decrease for most of the organic compounds studied. An increase in temperature generally decreases the absorption value. These values are obtained for long times (i.e., the "equilibrium" values).

From the absorption data, rubber particles can act as an effective absorber of a wide variety of organic liquid.[11]The particles can be mixed with other absorptive materials such as waste wood fines to increase absorption or absorption rate. It is most interesting to observe that if the rubber particles are placed on water, they do not sink but remain floating due to surface tension forces. This is especially true if the particles are milled to a 30 mesh, or smaller, size. Larger particles can contain fiber reinforcements which have considerably higher densities than water and subsequently force the particle to sink.

Oil spillage on water can easily be cleaned by casting particles upon the oil slick. Rapid oil absorption by the rubber particles ensues, creating a viscous floating mass which can easily be removed from the water surface by mechanical means.

A corollary method uses the rubber particles in convenient porous bags made of nonwoven polypropylene.[6] The bags can be floated along piers and shorelines to absorb the oil. These bags can also be useful for land spills of organic liquids such as gasoline at gas stations and airports or for vehicle floor mats to absorb dripping oils from the transmission, drive train, and engine.

It is interesting to note that pulverized rubber particles absorb dissolved mercury in water. This particular use of rubber particles as a heavy metal absorber is probably specific to those metals which easily form sulfides. The absorption greatly depends upon temperature and pH of water solution.

Recycled rubber can be utilized as a fuel.[8] In some applications small particles are not needed, and it is sufficient to utilize 5 to 7.6 cm (2 to 3 in.) pieces of rubber for full recovery. This has been done with coal mixture. It is curious that the rubber reduces sulfur emissions from the coal significantly. This is probably due to the reaction with ZnO compounds which produce nonvolatile ZnS compounds.

Particles of rubber can be destructively distilled or degraded to carbon blacks, oils, and lighter organics. The economics of carbon black production from rubber particles has recently become attractive due to the higher cost of benzene, which has been the traditional source for carbon black production.[8]

III. PLASTICS AND OTHER MATERIALS

The cryogenic recycling of other materials besides rubber has been under active investigation for several years. Some of these are listed in Table 5. The uses of these recycled materials are also indicated.

In the area of packaging films there are a large number of possibilities.[13] Most packaging films are composite laminates. Often cryogenic recycling is successful in separating an embrittled polymer from a more ductile polymer. Cooling can also embrittle the adhesive used in the laminate and aid in laminate separation. The separated polymer film can then be recylced back into the original extrusion operation or extruded for another purpose. Separation of foils from polymer coatings is especially aided by cryogenics.

Both thermoplastic and thermoset plastic wastes can be cryogenically recycled by grinding to the appropriate size. Thermoplastics can be re-extruded or molded, while thermosetting resins can be milled to a powder which is useful as an inexpensive filler material. This is especially useful for in-house recycling.

Electrical wiring and cabling can be readily recycled. The recovered metal (copper) is not greatly affected by the pulverization and can be sold to a metal fabricator for a premium price. The insulation can be cryogenically processed and reused as a binder, filler, or product.

Since the advent of the first symposium, Cryogenic Processing of Solid Wastes, held in 1976, numerous new applications of this straightforward processing method have been developed. The uses of these reprocessed materials not only depends upon the understanding of their properties but also upon the imagination and commitment of the scientist and engineer to recycling concepts.

REFERENCES

1. **Turner, J.,** President, private communication, Tirex International Research Foundation, Southlake, Tex., 1975.
2. Humane Equipment Corporation, private communication, Humane Equipment Corp., Baraboo, Wis., December, 1974.
3. Minnesota Mining and Manufacturing Corporation, private communication, Minnesota Mining and Manufacturing Corp., St. Paul, Minn., September, 1973.
4. Safe-T-Blade, Inc., private communication, Phoenix, Ariz., 1976.
5. **McDonald, C. H. and Olsen, R. E.,** Rubberized Asphalt Pavements, paper presented at 58th Annual Meeting, American Association of State Highway Officials, Phoenix, Ariz., November, 1972.
6. **Ziemendorf, D.,** Absorbent Bags for Oil Spill Clean-Up, Research Rep., Department of Mechanical Engineering, University of Wisconsin—Madison, 1979.
7. **Tharin, D. W.,** Dupont Report, 1974, DP-1343, Savannah River Laboratory, E. I. DuPont de Nemours, Aiken, S.C., 1974.
8. **Crane, G., Elefritz, R. A., Kay, E. L., and Laman, J. R.,** *Rubber Chem. Technol.,* 51, 3, 1978.
9. **Sanguandeekul, T.,** Properties of Molded, Recycled Rubber Particles and Their Composites, Master's of Science thesis, University of Wisconsin—Madison, 1975.
10. **Prasassarakich, P.,** A Study on Rubber Properties via Compounding and Compression Molding, Master's of Science thesis, University of Wisconsin—Madison, 1974.
11. **Schnoll, S. R.,** Absorption of Organic Compounds by Cryogenically Recycled Rubber Particles, Research Rep., Department of Chemical Engineering, University of Wisconsin—Madison, 1976.
12. **Klotz, D. and Clark, G.,** The Use of Recycled Tire Rubber Particles for Oil Spill Recovery, *Conservation and Recycling,* 1, 231, 1977.
13. **Koon Wing, Tam,** Recycling of Thermoplastic Films, Master's of Science thesis, University of Wisconsin—Madison, 1977.

Table 1
USES OF CRYOGENICALLY RECYCLED RUBBER PARTICLES

	Ref.
Molded or extruded products	
Hoses	1
Fence posts	1
Livestock mats	2
Artificial turf	3
Lawn mower cutting blades	4
Coatings, paints, sealants	
Waterproofing	
Acid proofing	
Roof coatings	
Vehicle undercoatings	
Crack sealants	
Highway asphaltics	5
Absorptive products	
Organic absorption bags	6
Mercury filter material	7
Fuels and products from selective degradation	
Mixtures with coal	8
Abrasive carbon blacks	
Oil	

Table 2
PHYSICAL PROPERTIES OF MOLDED RUBBER AT DIFFERENT PARTICLE SIZE, NATURAL RUBBER AND SBR[2]

Sample	Tensile strength pascals (psi)	Modulus of elasticity pascals (psi)	Maximum elongation (%)	Shore A hardness
18 mesh rubber[a]	4.96×10^5 (72.0)	$1.31 \pm .13 \times 10^6$ (190 ± 19)	75.0	49
30 mesh rubber[a]	7.96×10^5 (115.5)	$1.65 \pm .17 \times 10^6$ (240 ± 24)	110.9	56
50 mesh rubber[a]	5.94×10^5 (86.1)	$2.41 \pm .24 \times 10^6$ (350 ± 35)	80.0	65
50 mesh rubber[b]	7.69×10^5 (111.5)	$2.55 \pm .26 \times 10^6$ (370 ± 37)	80.0	65
Unfilled natural rubber	2.55×10^7 (3700.0)	$1.93 \pm .19 \times 10^6$ (280 ± 28)	730.0	47
Carbon black filled natural rubber	2.10×10^7 (3050.0)	$4.34 \pm .43 \times 10^6$ (630 ± 63)	520.0	60
Unfilled SBR	2.15×10^6 (312.4)	—	—	—
50 parts carbon black per 100 parts SBR	2.28×10^7 (3300.0)	$3.31 \pm .33 \times 10^6$ (480 ± 48)	540.0	60

[a] Speed of testing 0.508 cm/min (0.2 in./min).
[b] Speed of testing 1.27 cm/min (0.5 in./min).

Courtesy of Prasassarakich, P., A Study on Rubber Properties via Compounding and Compression Molding, Master's of Science thesis, University of Wisconsin—-Madison, 1974, 30.

Table 3
TYPICAL ABSORPTION VALUES OF RECYCLED RUBBER PARTICLES FOR VARIOUS ORGANIC LIQUIDS SUSPENDED ON OR IN WATER. (GMS ORGANIC LIQUID/GMS RUBBER PARTICLES)[11]

Mesh Size	T°F	T°C	−10, +20	−20, +30	−30, +40	−40, +50
Benzene	50°	10	1.7	2.1	2.5	2.7
	70°	21	2.1	2.4	2.7	3.3
	90°	32	2.8	3.0	3.3	3.7
$C\,Cl_4$	50°	10	3.3	4.0	5.4	8.0
	70°	21	3.3	4.1	5.3	7.8
	90°	32	2.2	2.7	3.4	4.3
White	50°	10	1.0	1.2	1.6	3.0
gasoline	70°	21	1.2	1.4	2.1	3.3
	90°	32	1.5	1.8	2.2	3.5
Xylene	50°	10	1.4	1.6	2.2	2.8
	70°	21	1.4	1.7	2.2	3.2
	90°	32	1.5	1.7	2.3	3.6
Ethylene	50°	10	1.5	1.7	1.9	2.1
dichloride	70°	21	1.6	1.7	2.0	2.4
	90°	32	2.0	2.3	2.6	2.8
Tetrachloro	50°	10	1.5	2.0	2.5	3.4
ethylene	70°	21	1.6	2.1	2.8	3.9
	90°	32	1.7	2.3	3.2	4.6
Tetrachloro	50°	10	1.6	2.0	3.0	3.8
ethane	70°	21	1.7	2.1	3.0	4.0
	90°	32	2.8	2.5	3.6	4.4
n-Butyl acetate (Saturated water solution)	70°	21	—	—	3.5	—

Note: Any residual insoluble organic liquid was detected by organic soluble dyes. Recycled rubber used was from passenger and truck tires. Adsorption values varied little from sample to sample. Values quoted here appear to be within 10% scatter.

Courtesy of Schnoll, S. R., Absorption of Organic Compounds by Cryogenically Recycled Rubber Particles, Research Report, Department of Chemical Engineering, University of Wisconsin—Madison, 1976, 5.

Table 4
TYPICAL OIL UP-TAKE VALUES[a] FOR VARIOUS OILS AND RUBBER PARTICLES MIXED WITH WOOD PINES[b]

21°C (70°F)

Rubber particle mesh size	Wood pines (%)	SAE 30 Distilled H_2O	SAE 40 Sea H_2O	SAE 40 Distilled H_2O	Crude Distilled H_2O	Crude Sea H_2O
20	0	1.8	1.8	1.8	0.5	0.6
20	25	—	—	—	0.8	0.8
20	50	—	1.8	2.8	1.1	1.1
20	75	—	—	—	1.4	1.6
20	100	3.3	—	3.3	1.8	1.8
30	0	2.0	—	1.8	0.8	0.8
30	25	—	—	—	—	—
30	50	—	—	1.8	—	—
30	75	—	—	—	—	—
30	100	3.3	—	3.3	1.8	1.8
45	0	2.2	2.3	2.1	0.9	0.9
45	25	—	—	2.3	1.1	1.1
45	50	—	2.4	2.3	1.4	1.4
45	75	—	—	—	1.6	1.6
45	100	3.3	—	3.3	1.8	1.8
70	0	2.3	2.8	3.0	1.6	1.7
70	25	3.1	—	—	—	—
70	50	3.7	2.8	—	—	—
70	75	—	—	—	—	—
70	100	3.3	—	3.3	1.8	1.8

[a] Weight oil/weight sorbent material.
[b] At 21°C (70°F).

From Klotz, D. and Clark, G., *Conservation and Recycling*, 1, 231, 1977.

Table 5
RECYCLING OF MATERIALS OTHER THAN RUBBER

Waste product	Ref.	Use of recycled material
Plastic packaging; films and foils	13	1. Film delaminates; separation of film; recycled in lower grade application 2. Trim scraps pulverized and recycled into extruder
Thermoplastic wastes; trim and flashing	14	Cryogenic grinding forms useable particulate material for extrusion or molding
Thermoset wastes	15	Cryogenic grinding forms cheap powdered filler additive
Electrical cabling and wiring	16	Separation of metal wire from insulation; wire material sold back to metal fabricator; insulation reextruded or burned

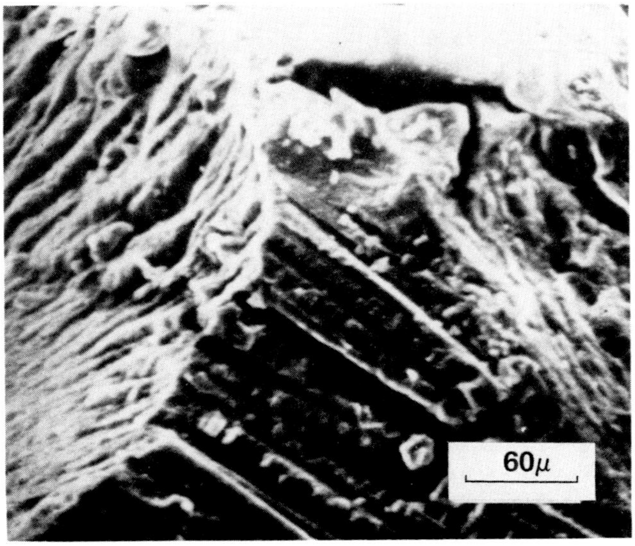

FIGURE 1. Surface of pulverized rubber particle. (Courtesy of Sanguandeekul, T., Properties of Molded, Recycled Rubber Particles and Their Composites, Master's of Science thesis, University of Wisconsin—Madison, 1975.)

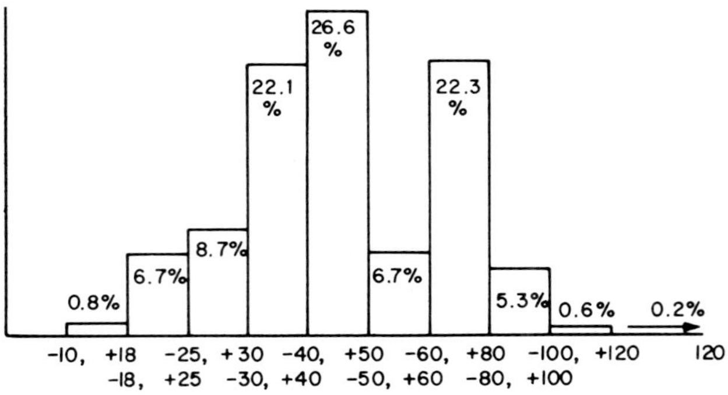

FIGURE 2. Size distribution of particulate rubber from pulverizer (high speed). (Courtesy of Sanguandeekul, T., Properties of Molded, Recycled Rubber Particles and Their Composites, Master's of Science thesis, University of Wisconsin — Madison, 1975.)

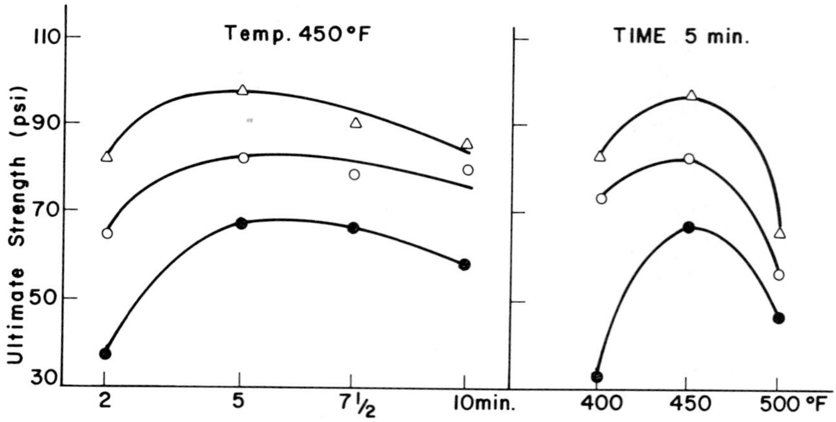

FIGURE 3. Ultimate strength vs. molding time and temperature, respectively. Molding pressure 180 psi. Cooling outside pressure. ●—10, + 18 mesh; ○—25, + 30 mesh; Δ—30, + 40 mesh. (Courtesy of Sanguandeekul, T., Properties of Molded, Recycled Rubber Particles and Their Composites, Master's of Science thesis, University of Wisconsin — Madison, 1975.)

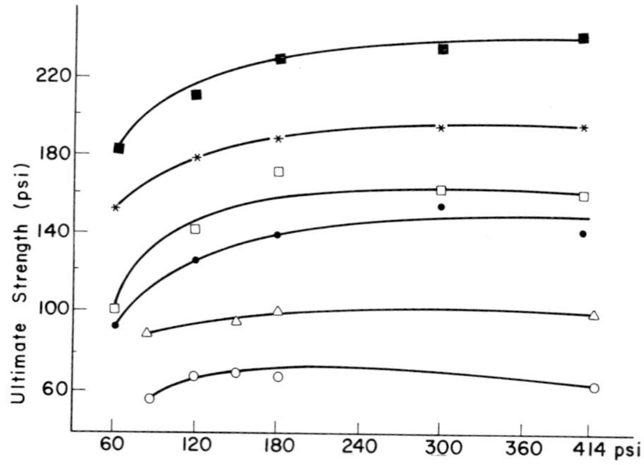

FIGURE 4. Ultimate strength vs. molding pressure. ○—10, + 18 mesh; Δ—18, + 30 mesh; ●—25, + 30 mesh; □—30, + 40 mesh; *—40, + 50 mesh, ■—60, + 80 mesh. Molding temperature 450°F; molding time 5 min. (Courtesy of Sanguandeekul, T., Properties of Molded, Recycled Rubber Particles and Their Composites, Master's of Science thesis, University of Wisconsin — Madison, 1975.) Compression molding, Master's of Science thesis, University of Wisconsin — Madison, 1974.)

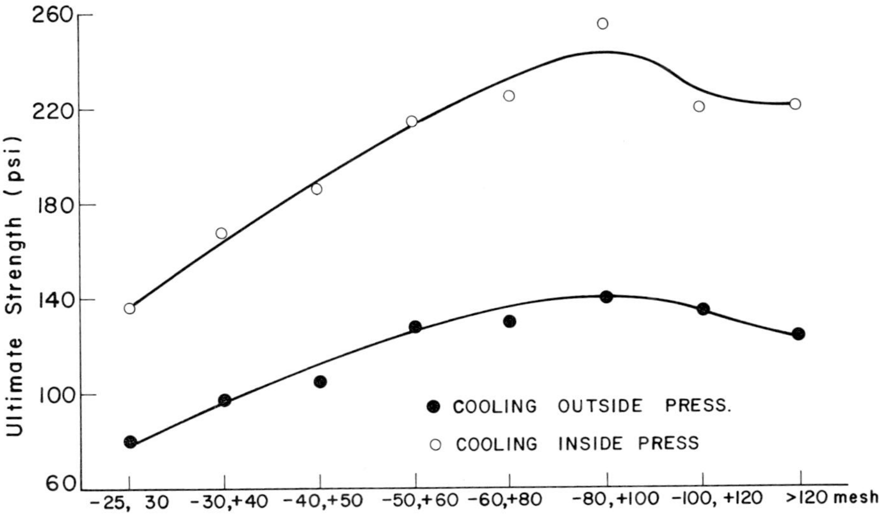

FIGURE 5. Ultimate strength vs. particle size of molded rubber. Molding temperature 450°F, molding time 5 min; molding pressure 180 psi. (Courtesy of Sanguandeekul, T., Properties of Molded, Recycled Rubber Particles and Their Composites, Master of Science thesis, University of Wisconsin — Madison, 1975.)

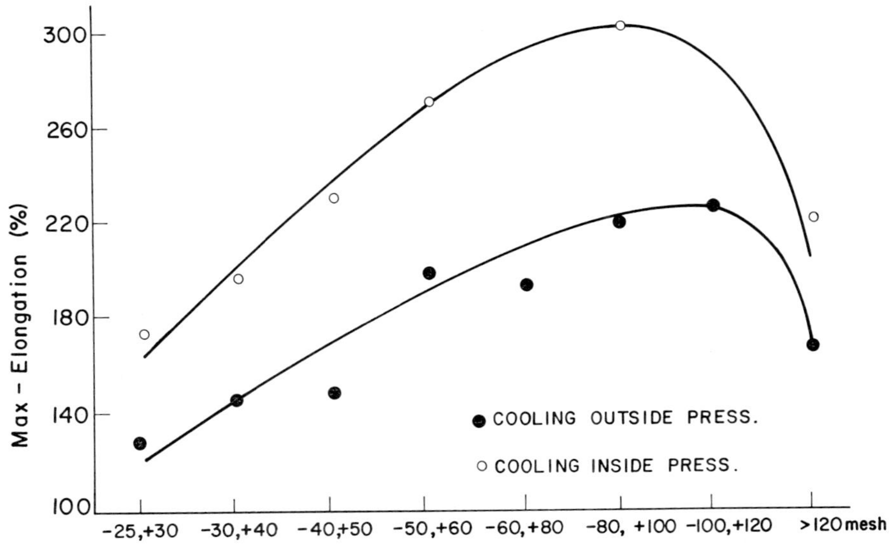

FIGURE 6. Max-elongation vs. particle size of molded rubber. Molding temperature 450°F; molding time 5 min; molding pressure 180 psi. (Courtesy of Sanguandeekul, T., Properties of Molded, Recycled Rubber Particles and Their Composites, Master's of Science thesis, University of Wisconsin — Madison, 1975.)

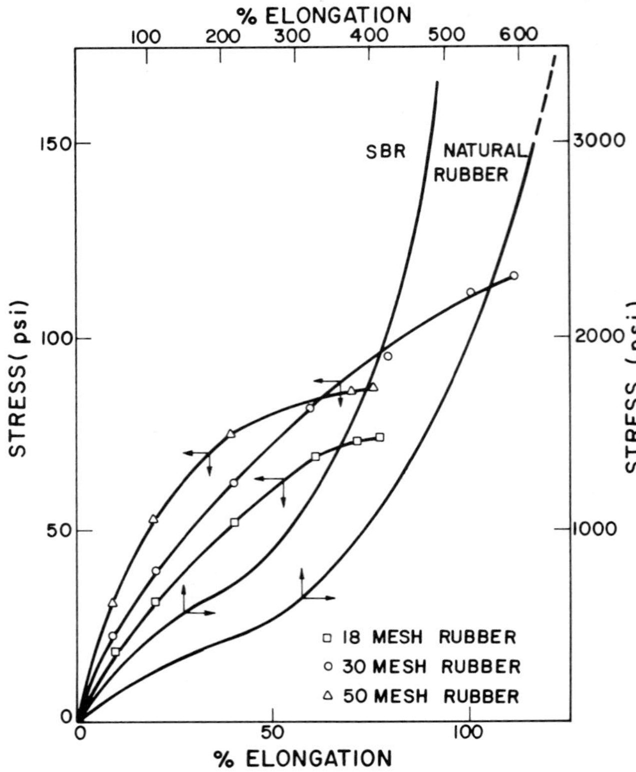

FIGURE 7. Stress-strain curves of molded rubber, of sizes 18, 30, and 50 mesh at strain rate 0.2 in./min unfilled natural rubber and 50 phr carbon black filled SBR. (Courtesy of Sanguandeekul, T., Properties of Molded, Recycled Rubber Particles and Their Composites, Master's of Science thesis, University of Wisconsin — Madison, 1975.)

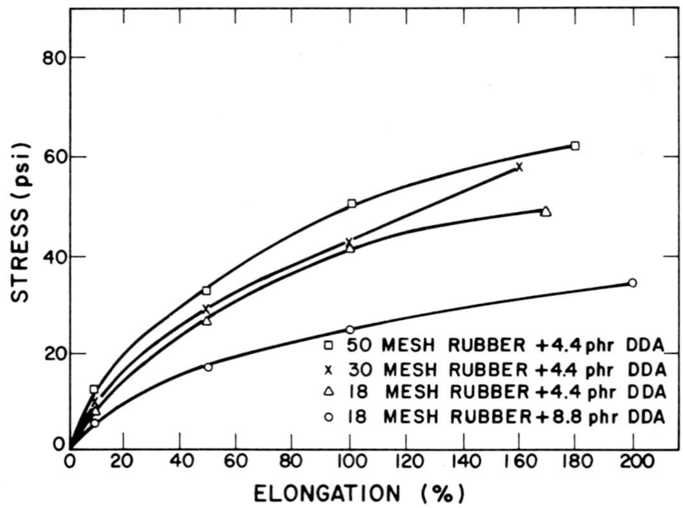

FIGURE 8. Stress-strain curves for molded reclaimed rubber. (Courtesy of Sanguandeekul, T., Properties of Molded, Recycled Rubber Particles and Their Composites, Master's of Science thesis, University of Wisconsin — Madison, 1975.)

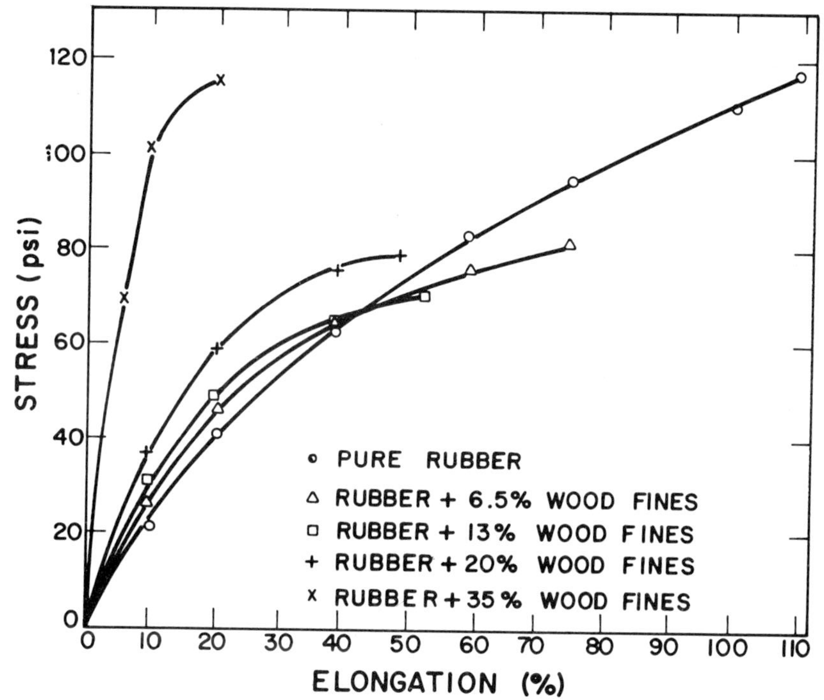

FIGURE 9. Stress-strain curves for rubber (30 mesh) mixed with wood fines. (Courtesy of Sanguandeekul, T., Properties of Molded, Recycled Rubber Particles and Their Composites, Master's of Science thesis, Univerity of Wisconsin — Madison, 1975.)

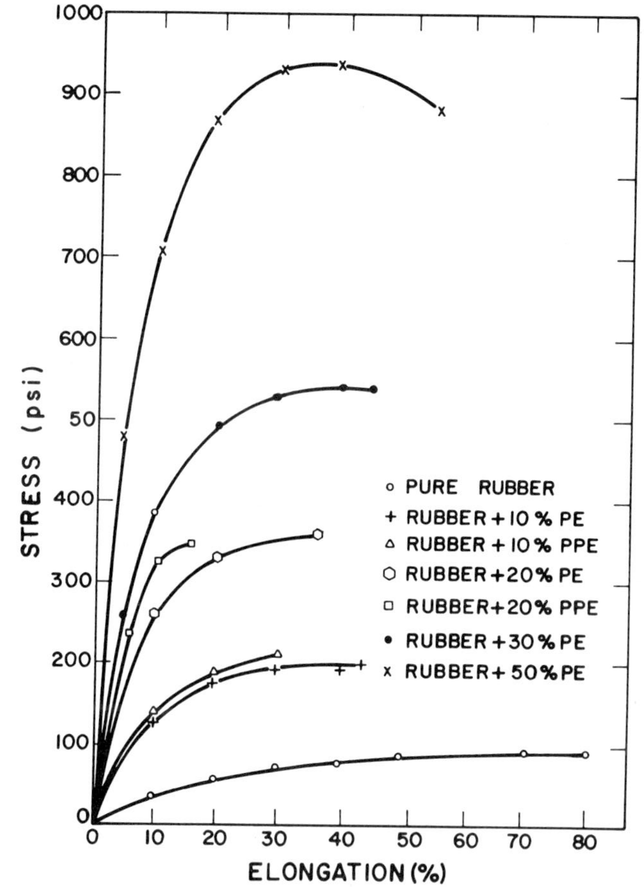

FIGURE 10. Stress-strain curves for rubber (50 mesh) with PE and PPE. (Courtesy of Prasassarakich, P., A study of Rubber Properties via Compounding and Compression Molding, Master's of Science thesis, University of Wisconsin — Madison, 1974.)

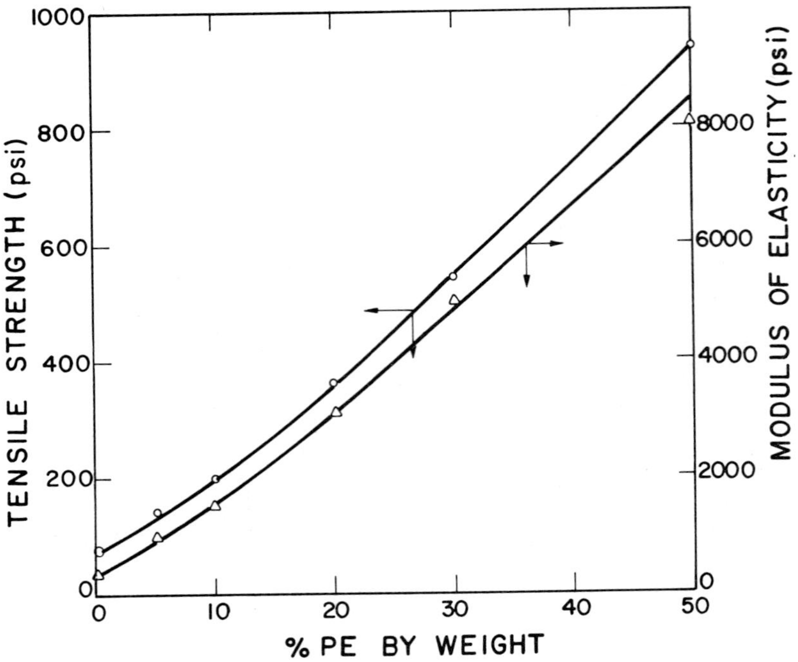

FIGURE 11. Effect of amount of PE on modulus of elasticity and tensile strength of rubber. (Courtesy of Prasassarakich, P., A Study on Rubber Properties via Compounding and Compression Molding, Master's of Science thesis, University of Wiisconsin — Madison, 1974.)

Chapter 9

CRYOGENIC PROCESSING OF FOODSTUFFS

J. M. Venetucci

TABLE OF CONTENTS

I. INTRODUCTION

Freezing has long been recognized as an excellent method of food preservation. Since early times, farmers, fishermen, and trappers residing in cold weather climates preserved their fish and game by freezing and storing them in unheated buildings. Even commercial freezing is not new. In 1865, during the cold season, fishermen of the Great Lakes would allow their catch to remain on the ice several hours for freezing. This method is referred to as "weather freezing" and is still used when weather permits. Weather-frozen products produced good profits because of their high quality.

What is known as the "frozen food industry" emerged when "fresh frozen" products appeared on the market in the early 1930s. The industry's growth was slow until the postwar era when household refrigeration and freezer lockers became readily available. Retail store owners generally agree that the best items to sell are frozen food products. Dollar sales returned per square foot of floor space and gross profit margin are considerably higher for frozen foods than for other items.

The acceptance of frozen foods by the consumer can be partially attributed to the convenience of fast and quick preparation in the home as well as in the "fast food" type of restaurants. The ability to retain quality by freezing has expanded the process to include entire entrées. The production increase in frozen products began in 1945. Now demand for frozen food products can be expected to double every 10 years.

This dramatic and steady increase is partially due to the convenience nature of frozen foods and also to the major factors of stability and maintained quality. Stability can be defined as "steadiness" or the ability to endure without substantial change from the original. The length of time of consideration is referred to as the "stable period".

Quality cannot be defined with one word, but must be decided upon by personal acceptance. Normally, trained observers note the flavor, appearance, smell, and texture of many samples. After some length of time, these observers will generally agree that the samples are distinguishably different in flavor, appearance, smell, and texture from the original.

It is generally agreed upon that rapid freezing of food is the best-known way to maintain quality and stability. In later sections these points will be discussed in detail, thus enabling us to effectively evaluate the benefits attainable from cryogenic freezing.

II. CRYOGENIC FOOD FREEZING TECHNOLOGY

Freezing of food is a basic heat transfer process where heat is extracted for the purpose of cooling the food and solidifying the moisture present.

Acid constituents and dissolved solids cause the freezing point of most foods to be lower than the commonly referred to 0°C (32°F) freezing point of pure water. As first ice crystal formation begins, usually in the −2.2°C (28°F) to −1.1°C (30°F) temperature range, a concentration of the dissolved constituents is produced, thereby reducing the freezing point further. If the freezing process is significantly slow, actual concentration gradients are established, and the resultant freezing point temperature range is from −37.22°C (−35°F) to approximately −1.1°C (30°F), with the most common being about −6.6°C (20°F). The size of the ice crystals formed within the food product is a function of the speed of freezing. Large ice crystals produced by slow freezing techniques (as shown in Figure 1) develop intercellular spaces, creating tension and increasing the tendency for the crystals to rupture cell walls. This allows cellular fluids to migrate during thawing. This is known as "drip loss" or "weep", and carries away dissolved solids, flavor constituents, aroma constituents, and nutrients, and significantly alters the texture of the food products. Figure 2 illustrates the possible weepage effect and loss of quality of frozen strawberries.

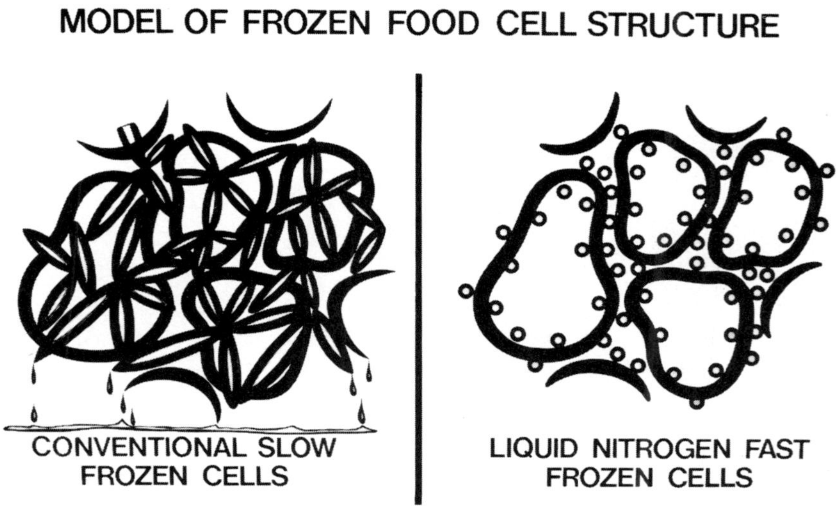

FIGURE 1. Model photograph of frozen food cell structure. (Courtesy of Liquid Carbonic Corporation, Chicago.)

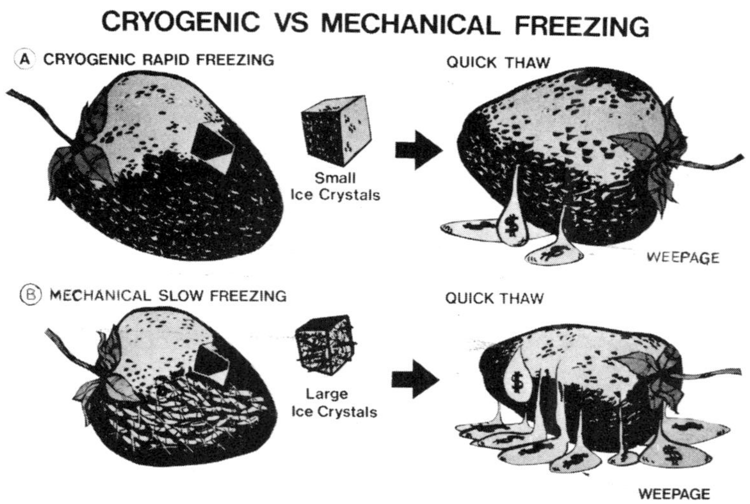

FIGURE 2. Cryogenic vs. mechanical freezing and weepage. (Courtesy of Liquid Carbonic Corporation, Chicago.)

Fast freezing rates at cryogenic temperatures offer many advantages to the frozen food industry. These advantages led to the investigation and adaptation of using an expendable liquefied gas such as liquid nitrogen to be used as a freezing medium. For comparison, other conventional freezing systems — of the plate type or the blast freezer design — are generally limited to process temperatures of −23.33°C (−10°F) to −51.11°C (−60°F). Liquid nitrogen however, has an extremely low temperature of −195.8°C (−320.4°F), and can be used to produce any desired range of freezing temperatures.

As previously stated, product quality and flavor retention are directly related to the size and formation of ice crystals. Hence, the faster the freezing rate, the smaller the ice crystal. Figure 3 illustrates one of the major benefits of cryogenic freezing. Identical

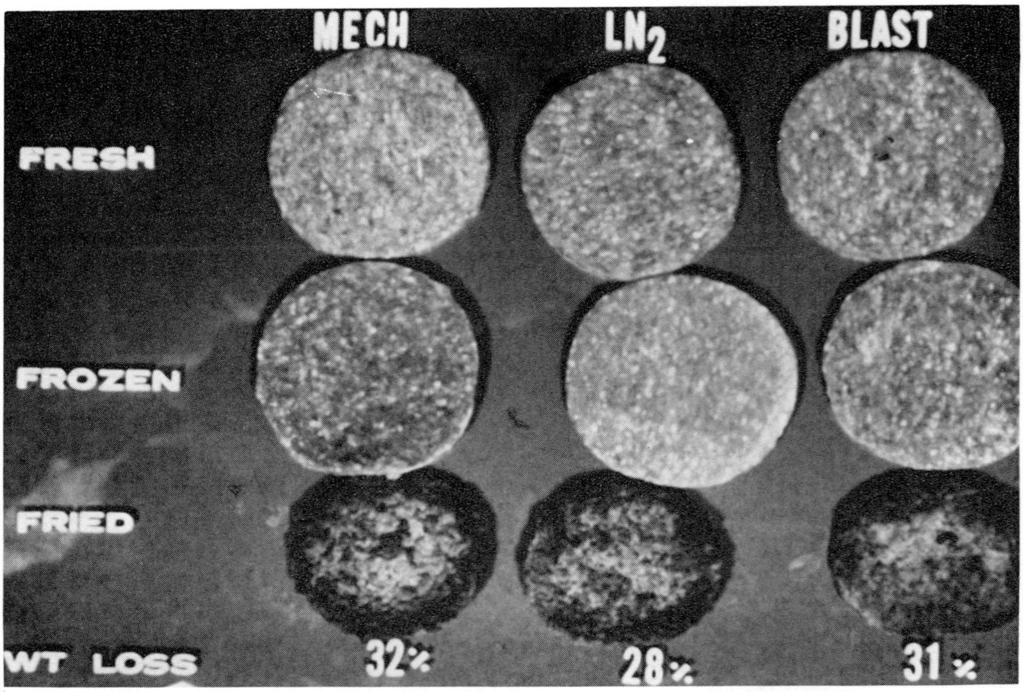

FIGURE 3. Comparison of typical weight-loss after identical cooking — mechnical-, blast, and liquid nitrogen-frozen 4:1 (4 patties per pound) hamburger patties. (Courtesy of Liquid Carbonic Corporation, Chicago.)

hamburger meat patties were cooked for the same period of time and on the same grill. The cryogenically frozen patty exhibited less weight loss than the blast or mechanically frozen patty.

A. Rate of Freezing Effects

1. Ice Crystal Location

It is apparent that the consequences that result from slow or rapid freezing be reviewed since the rate of freezing does influence quality.

During slow freezing, the food product remains close to the initial freezing point for a prolonged period of time. During this time period, crystal growth is active. The nuclei which form generally grow extensively in extracellular locations. Extracellular crystal growth at the freezing region is irreversible and is due to osmotic pressure. This water immigration forms large crystals within the intercellular regions, thereby forcing fiber shrinkage. Slow freezing provides an opportunity for water to migrate and results in frozen tissue which has a distorted appearance while in the frozen state.

During rapid freezing, however, the food product is cooled rapidly to temperatures below the initial freezing point. The water crystal nuclei, therefore, have a tendency to form uniformly throughout the cell and grow to only a limited size because of the vast number of nuclei present. Rapid freezing decreases the possibility for osmotic pressure and migration and results in interrupted freezing boundaries. Products that are rapidly frozen, therefore, will generally have normal appearance. Furthermore, minimal damage to texture and loss of fluid during thawing will occur.

2. Cell Damage

The rate of freezing, in general, will not have any significant influence on the volume

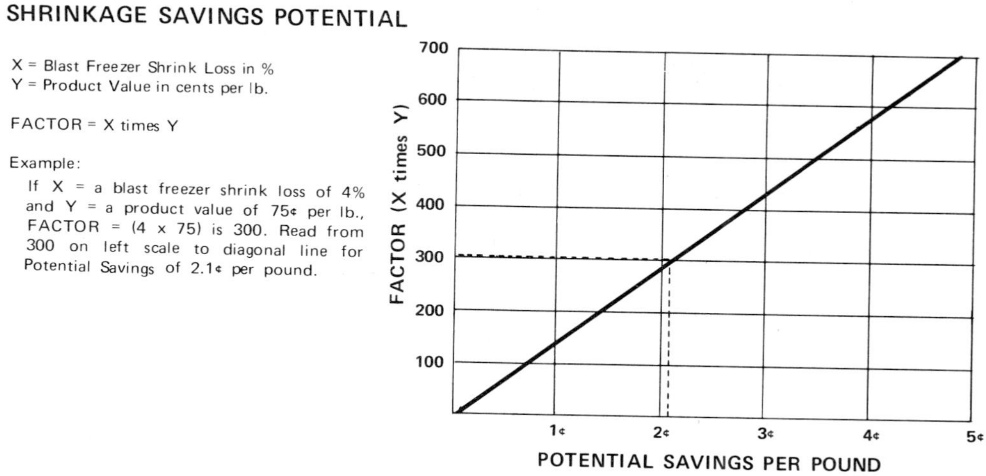

FIGURE 4. Shrinkage savings potential. (Courtesy of Liquid Carbonic Corporation, Chicago.)

change that occurs during freezing. Animal tissue is basically free of air. As previously mentioned, large ice crystals are produced by slow freezing. These large crystals will expand while the tissue contracts. It is obvious, therefore, that areas within these cells will contract and expand and produce ruptured cells. This undesirable condition is not as apparent in rapid freezing.

3. Chemical Damage

In slow freezing, the oils, salts, and acids tend to concentrate in the large ice crystals and affect the eutectic point. In contrast, rapid freezing favors small ice crystals that entrain the concentrates and minimizes eutectic changes.

With these factors being considered, it would appear that rapid freezing minimizes freezing damage and maintains "stability" and quality. Additional advantages of rapid freezing will be investigated in later sections.

B. Dehydration

Dehydration loss with liquid nitrogen cryogenic freezing systems is insignificant. In some cases, it will be as low as 0.1% by weight. This compares with mechanical air-blast freezing losses in the range of 3 to 6% depending on the product. This reduction of shrink loss alone can often compensate for the total cost of cryogenic freezing. The higher the value of the food, the more dramatic are these savings (Figure 4).

Liquid nitrogen that is vaporized by the exchange of heat with the warmer food can absorb and carry away only the amount of moisture which is required to saturate the nitrogen at the temperature at which it leaves the cryogenic freezing chamber.

In addition to the higher product yield, the unique nature of cryogenic freezing results in improved product texture, taste, and appearance; the latter becoming increasingly important since a large percentage of frozen products are being sold to the consumer market through clear packages.

C. Cryogenic Freezing Methods

There are basically three methods of liquid nitrogen freezing: immersion freezing, spray freezing, and cooled-air freezing. In the immersion-freezing process, the product is actually dipped into the liquid nitrogen. In this process, usually only the latent heat of the liquid nitrogen is utilized.

The spray-freezing process is similar to immersion-cooling in that the product actually comes in contact with liquid nitrogen. With spray freezing, however, the liquid nitrogen is sprayed onto the product from positioned nozzles. When the liquid nitrogen comes in contact with the product, it absorbs heat and transforms into a gas vapor. Some of the sensible heat still remaining in the cold gas vapor can be utilized by precooling the incoming product.

In the cooled air-freezing process, the product is frozen by means of cold air currents passing over it — the air being cooled by liquid nitrogen. This method is not very economical.

Initial experiments with liquid nitrogen immersion techniques showed that compared with conventional methods, the time to freeze a given product was drastically reduced and the taste of the thawed product substantially improved. However, initially three problems arose:

1. Lack of efficient contact between the product and the liquid nitrogen caused by violent boiling and gas-film formation at the product surface
2. Cracking and peeling of many products caused by tremendous thermal shock and violent contraction of the outer surface
3. Inefficient use of the refrigeration available in the nitrogen vapor

It quickly became clear that more sophisticated techniques were required to eliminate the problems. It was found that the use of the cold nitrogen vapor to precool a product prevented surface cracking of the food and violent boiling of liquid nitrogen. Cooling the product with the cold gas first results in gradual contraction of the surface and prepares the product for contact with the liquid nitrogen. The cooler product surface also reduces the violent boiling action that occurs when a warm product is immersed in liquid nitrogen. Spraying the surface with liquid nitrogen in a cryogenic environment further eliminates the violent boiling, while providing an even greater heat-transfer rate than that achieved by the immersion technique. As a result of the perfected nitrogen flash-freezing technique, a wide range of new frozen food products are now successfully processed and marketed. These products, as shown in Figure 5, include tomatoes, asparagus tips, sliced peaches and pears, strawberries, egg rolls, pizza, seafoods, dairy products, bakery goods, meats, poultry, and precooked foods. (A case study on liquid nitrogen tomato peeling is presented in Chapter 10.)

The cryogenic freezing process is especially suited to high moisture content foods, since large, jagged ice crystals do not form and puncture food cell walls. Natural juices are retained and food discoloration, moisture, and vitamin loss are minimized or completely eliminated. Several examples of cryogenically processed food products are shown in Figure 6. Cryogenic freezing offers the frozen food processor the following advantages over other conventional freezing methods.

1. **Improved Product Quality** — An immediate crust-freeze locks in flavor and quality of the product.

2. **New Products** — Liquid nitrogen allows the processing of many products once considered impossible to freeze — e.g., tomatoes, avocados, mushrooms, and whole strawberries.

3. **Increased Production** — Production rates can be substantially increased without the need for additional floor space.

4. **Greater Economy** — Efficient design minimizes liquid nitrogen consumption. Capital investments for equipment are low compared to costs of other conventional freezing equipment of equal capacity.

FIGURE 5. Typical products frozen with liquid nitrogen. (Courtesy of Liquid Carbonic Corporation, Chicago.)

5. Minimum Dehydration — Cryogenic freezing substantially reduces the amount of dehydration of the food product compared to conventional freezing.

D. Cryogenic Freezing Tunnels

There are numerous types of cryogenic freezing tunnels presently available. Each manufacturer, however, has incorporated specific features in an effort to increase the efficiency of the individual tunnel. There is, however, one common feature in the majority of tunnels — that is, the conveyance of the food on a belt through a low-temperature atmosphere in which heat is removed from the food with the use of a cryogenic liquid. The efficiency of the machine will be directly related to the manner in which the cryogen is dispensed and the amount of refrigeration that is effectively used from the latent heat and sensible heat of the cryogen.

A typical liquid nitrogen freezing tunnel is shown in Figure 7, and its major components are illustrated in schematic form in Figure 8. The description of the operation of the freezer is described below.

The basic arrangement of the machine consists of a product conveyor mounted within a double wall, highly insulated, open-ended tunnel. A liquid nitrogen spray zone is located in a section near the exist end of the unit and includes spray nozzles above the product. A blower fan for recirculation of the cold gas is mounted adjacent to the tunnel. Ducting directs the cold gas over the surface of the product. The gas recirculating zone is located between the entrance end and spray zone of the chamber and occupies the greater portion of the length of unit. Fans are located along the gas recirculating zone to add additional turbulence to the nitrogen gas, thus increasing heat transfer rate from the product.

A

FIGURE 6. Examples of liquid nitrogen-processed food products. (A) Preformed and breaded onion rings prior to entering a cryogenic freezing tunnel; (B) Portion control meat products such as veal steaks after being breaded and about to enter a cryogenic freezing tunnel; (C) Specialty foods that are precooked such as Cornish hens shown as they enter a freezing tunnel. (Courtesy of Liquid Carbonic Corporation, Chicago.)

B

C

FIGURE 7. Typical liquid nitrogen freezing tunnel. (Courtesy of Hollymatic Corporation, Park Forest, Ill.)

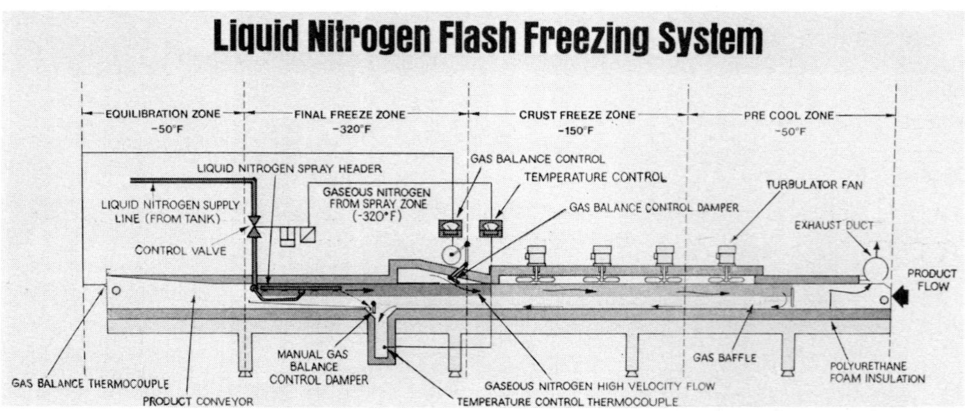

FIGURE 8. Schematic of a freezing tunnel. (Courtesy of Hollymatic Corporation, Park Forest, Ill.)

Single or dual conveyors within the tunnel are used to move the product into and out of the cryogenic atmosphere. The conveyor speed is adjustable in order to vary the exposure time of the product within the freezer.

There are several methods of supplying liquid nitrogen to a cryogenic freezer. In the system described, liquid nitrogen from the storage vessel is supplied to a control valve located at the freezer and then to a spray header directing liquid nitrogen against the product being frozen. Temperature in the machine gas zone is controlled by this spray header valve modulation.

Excess liquid which has not flashed to gas when in contact with the product and liquid which does not come in contact with the product is collected in a pan under the spray zone and flows to a point in the gas recirculation zone. The liquid is then mixed with recirculated cold nitrogen gas vapors by means of the recirculation blower.

In effect, the heat load of the product entering the freezer automatically determines the demand for liquid nitrogen from the supply tank.

The economic feasibility of freezing with liquid nitrogen largely depends on the amount of heat absorption that can be gained by recirculating the −195.8°C (−320.4°F) nitrogen gas vapor which is generated in the liquid spray zone of the freezer. For the change of state (liquid to gas), 86 Btu/lb of liquid nitrogen is required, and an additional 0.25 Btu/lb is required for each degree of temperature increase of the gas. In order to take advantage of the heat absorption of the cold gas, a recirculation system is used to move the gas at high velocity over the surface of the product. Additional turbulator fans are added along the gas recirculation path to gain added heat transfer between the product and colder gases. The amount of gas exiting from both ends of the tunnel is automatically controlled so that the greater portion is exhausted out of the entrance end of the freezer.

This, in effect, subjects the warmest product to the warmest gas. As the product progresses through the freezer, it encounters colder gas temperatures until finally it passes through the spray environmental zone maintained at −195.8°C (−320.4°F0). This is analogous to a counterflow heat exchanger which is thermally more efficient than parallel flow.

E. Advantages of Liquid Nitrogen Freezing Equipment

There are several advantages available when cryogenic freezing methods are used. A comparison between liquid nitrogen and conventional mechanical freezing methods is shown in Table 1.

Food processing plants make use of all available floor space for processing, with very little floor area available for large mechanical equipment such as mechanical refrigeration compressors, etc. Together with blast freezing rooms, a simple and comparative illustration of the equipment required for each system is shown in Figure 9.

F. Cooling and Freezing Calculations

It may be advantageous to understand the simple calculations that are used to determine the economical efficiencies of considering the use of liquid nitrogen for cryogenic freezing.

The process is one of two or more stages of refrigeration where simple cooling is used to lower the temperature of a substance to its freezing point where it is supplied with sufficient refrigeration to freeze the product with no further reduction in temperature. In some cases, the frozen product may undergo subsequent simple cooling with corresponding temperature change in order to reach a desired storage temperature. The process may be broken down into steps of cooling, freezing, and subcooling as shown below.

$$Q = W C_1 (T_1 - T_f) + W X + W C_2 (T_f - T_2) \qquad (1)$$

Total refrigeration = cooling + freezing + subcooling.

Where Q = refrigeration required (Btus); W = weight of substance (lbs); C_1 = specific heat above freezing point (Btu/lb °F); C_2 = specific heat below freezing point (Btu/lb °F); T_1 = initial temperature (°F); T_f = freezing point of meat (°F); T_2 = final temperature of meat (°F) X = latent heat of freezing of product (Btu/lb).

Example — A theoretical problem might require the cooling and freezing of 100 lb of hamburger patties. The meat has an initial temperature of 1.7°C (35°F) and it is required to reduce the temperature of the meat to −17.8°C (0°F) prior to placing it into storage. From tables, we can obtain the specific heat value of the meat above freezing and the specific heat value of the meat below its freezing temperature.

Table 1
ADVANTAGES OF LIQUID NITROGEN FREEZING OVER CONVENTIONAL METHODS

Liquid nitrogen freezing	Conventional freezing
Freezer system installs compactly and economically	System requires use of costly space consuming equipment
Nitrogen freezing virtually eliminates cell damage, retains nearly 100% food moisture; maintains natural drip, flavor, color, and quality of products; disrupts bacterial growth due to extremely rapid freezing in an inert nitrogen environment	Slow freezing ruptures food cells due to formation of large ice crystals within the cells; when thawed or cooked, product experiences drip loss; flavor, appearance, nutrition, and shelf life are lowered; an additional 1½ to 7% dehydration loss occurs with slow freezing
Pack-off product handling greatly reduced when an in-line nitrogen freezer is installed between processing machines; nitrogen freezing offers greatest production rates per square foot due to minimal freezing duration	Conventional freezing requires a greater fixed amount of space because the product must remain in the freezer longer to reach the same temperature acquired quickly in the nitrogen freezer; since conventional freezers are seldom in line, handling in and out creates a bottleneck
Nitrogen freezers can be cleaned in minutes, not hours, and easily surpass all stringent USDA regulations; since a nitrogen freezer is only operated during production hours, cleaning is done in comfortable room temperature; no need for arctic clothing	Cleaning a conventional freezer is time-consuming; panels, frames, rollers, and long belts are difficult to reach; varying component heights, cold temperature, and inaccessible areas create worker inefficiency
The turnup of the nitrogen freezer and its expandability simplifies increased production	The conventional freezer is fixed in place, with all its support equipment; extremely difficult to expand, impractical to move, and its capacity is fixed
Nitrogen freezers are relatively maintenance-free; piping and storage vessels are completely maintenance-free; a nitrogen freezer requires one nitrogen line, one electrical drop, and an exhaust line; simple and dependable	It takes a great deal of equipment to make a conventional freezer work and more ability to keep it running; compressors, condensers, evaporative towers, and extensive plumbing contribute to mechanical problems and expensive downtime

In this example, we will assume a specific heat value of 0.8 Btu/lb °F for above freezing and value of 0.5 Btu/lb °F for below freezing. The latent heat of freezing the meat from 1.7°C (35°F) to −17.8°C. (0°F) is 115 Btu/lb and is determined from specific heat curves for food similar to the one shown in Figure 10. The freezing point of the meat will be assumed to be −6.7°C (20°F). It must be pointed out that the specific heat of the meat will be directly related to the moisture and fat content. Therefore, every different type of hamburger blend would have different specific heat values.

We are required to determine the amount of refrigeration in Btus that is needed, and the liquid nitrogen in pounds to freeze the 100 lb of meat. Figure 11 illustrates what a Btu is, and its quantity or value is the same for heat input or removal from a product.

We have 166 Btu/lb available from one pound of liquid nitrogen. In this example, assume a 70% efficiency factor for the cooling process.

Substituting in our equation:

$$Q = WC_1(T_1 - T_f) + WX + WC_2(T_f - T_2)$$

$$Q = W[C_1(T_1 - T_f) + X + C_2(T_f - T_2)]$$

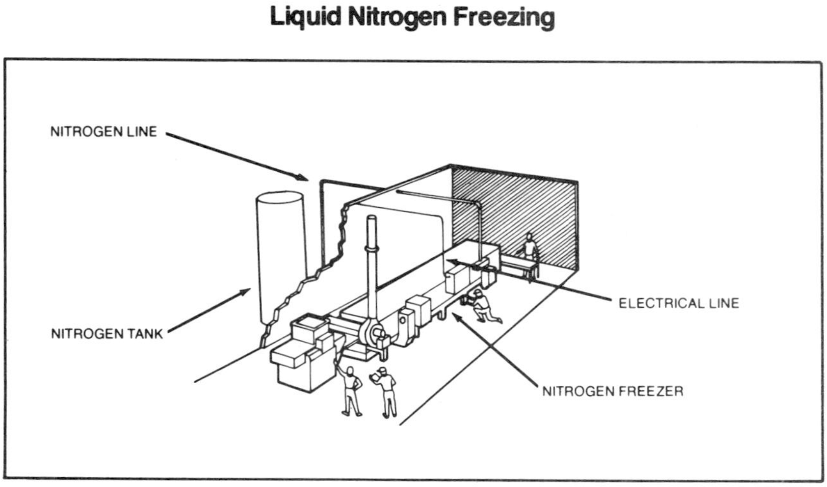

A

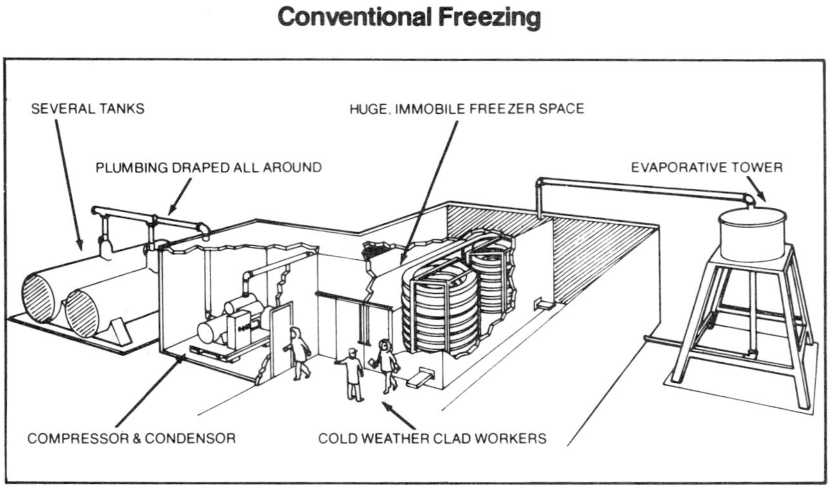

B

FIGURE 9. Liquid nitrogen freezing (A) vs. conventional freezing equipment (B). (Courtesy of Hollymatic Corporation, Park Forest, Ill.)

$$
\begin{aligned}
Q\ \text{Btu} &= 100\ \text{lb}\ [(0.8\ \text{Btu/lb}\ ^\circ\text{F}\ (35^\circ\text{F} - 20^\circ\text{F}) + 115\ \text{Btu/lb} + 0.5\ \text{Btu/lb}\ ^\circ\text{F}\ (20^\circ\text{F} - 0^\circ\text{F})] \\
Q\ \text{Btu} &= 100\ \text{lb}\ [(0.8\ \text{Btu/lb}\ ^\circ\text{F}\ (15^\circ\text{F}) + 115\ \text{Btu/lb} + 0.5\ \text{Btu/lb}\ ^\circ\text{F}\ (20^\circ\text{F})] \\
Q\ \text{Btu} &= 100\ \text{lb}\ [(12\ \text{Btu/lb} + 115\ \text{Btu/lb} + 10\ \text{Btu/lb})] \\
Q\ \text{Btu} &= 100\ \text{lb}\ (137\ \text{Btu/lb}) \\
Q\ \text{Btu} &= 13{,}700\ \text{Btu}
\end{aligned}
$$

Solving for the amount of liquid nitrogen required

$$
W_{LN_2} = \frac{13{,}700\ \text{Btu}}{166\ \text{Btu/lb (Liquid } N_2) \times 70\%} \qquad (2)
$$

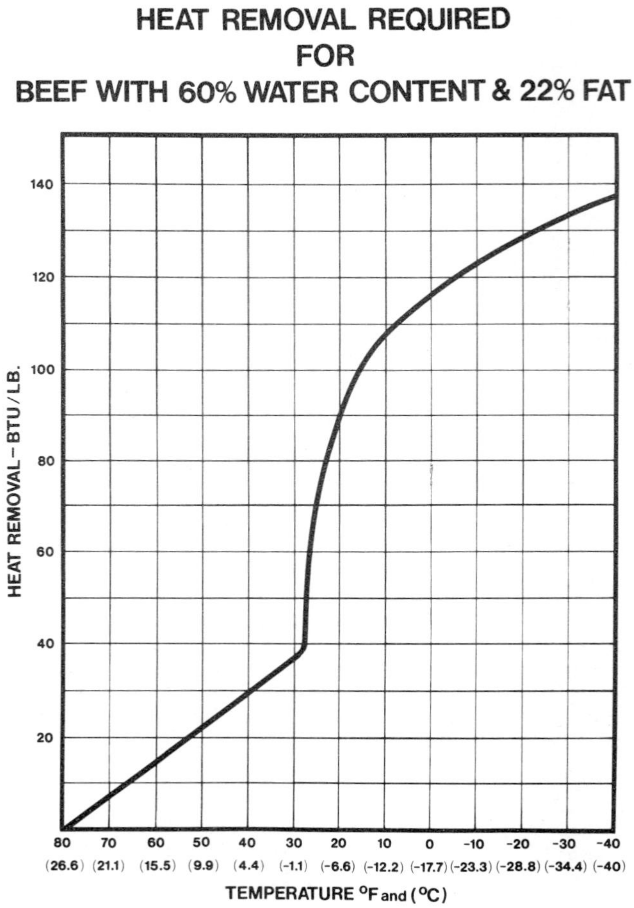

FIGURE 10. Heat removal required for beef. (Courtesy of Liquid Carbonic Corporation, Chicago.)

$$W_{LN_2} = 118 \text{ lb} \tag{3}$$

Similar calculations can be made for other food products and in the same manner. A tabulation of the specific heat values of many of the more common foods are shown in Table 2.

The basic concept of simple cooling and freezing can be expanded to include simple calculations to determine the approximate size of a liquid nitrogen freezing tunnel together with the amount of liquid nitrogen required for freezing.

Example — Determine the size of a freezer that would be required to freeze hamburger patties. The patties would weigh 4 oz each and 14 cm (5.5 in.) in diameter. They would be loaded on a conveyor 96.5 cm (38 in.) wide with six patties across the belt. The production rate is 1220 kg/hr (2700 lb/hr) for 10 hr/day. Initial meat temperature is 4.4°C (40°F) and a desired final temperature is −17.78°C. (0°F). The moisture content of the meat is established to be 60%.

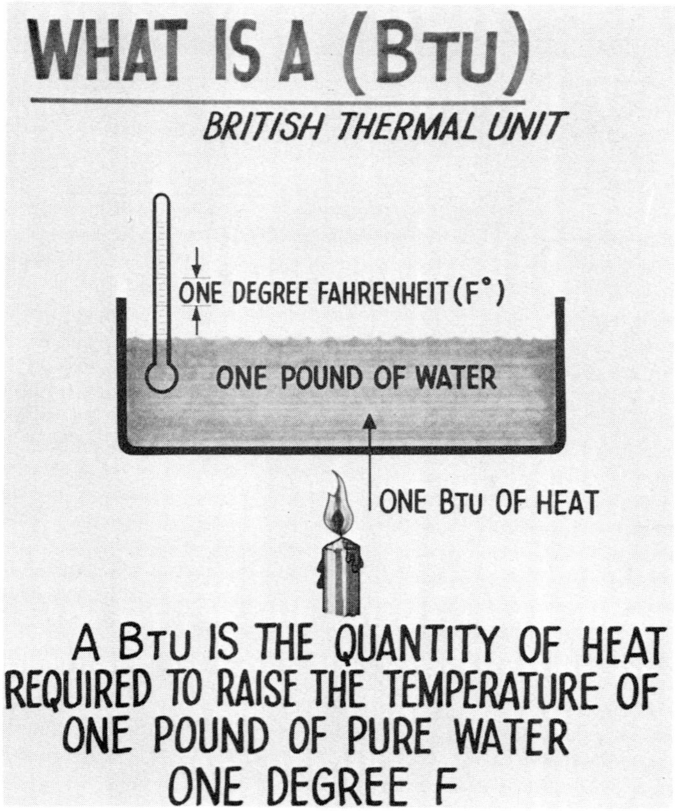

FIGURE 11. What is a Btu. (Courtesy of Liquid Carbonic Corporation, Chicago.)

Table 2
SPECIFIC HEAT VALUES OF COMMON FOODS

Product	Average freezing point		Water %	Specific heat (Btu/lb °F)		Latent heat of fusion (Btu/lb)
	°C	°F		Above freezing	Below freezing	
			Fruits			
Apples	−2.2	28	84	0.86	0.45	121
Apricots	−2.2	28	85	0.88	0.46	122
Avocados	−2.7	27	94	0.91	0.49	136
Bananas	−2.2	28	75	0.80	0.42	108
Blackberries	−1.6	29	85	0.88	0.46	122
Blueberries	−1.6	29	82	0.86	0.45	118
Cantaloupes	−1.6	29	93	0.94	0.48	132
Cherries	−3.3	26	83	0.87	0.45	120
Cranberries	−2.7	27	87	0.90	0.46	124
Currants	−1.1	30	85	0.88	0.45	120
Dates	−2.7	27	78	0.82	0.43	112
Figs	−2.7	27	78	0.82	0.43	112
Gooseberries	−1.6	29	88	0.90	0.46	126
Grapefruit	−2.2	28	89	0.91	0.46	126
Grapes	−3.3	26	82	0.86	0.44	116

Table 2 (continued)
SPECIFIC HEAT VALUES OF COMMON FOODS

Product	Average freezing point °C	Average freezing point °F	Water %	Specific heat (Btu/lb °F) Above freezing	Specific heat (Btu/lb °F) Below freezing	Latent heat of fusion (Btu/lb)
Honeydew melon	−6.6	20	93	0.94	0.48	132
Lemons	−2.2	28	89	0.92	0.46	127
Limes	−1.6	29	86	0.89	0.46	122
Mangoes	0.0	32	93	0.90	0.46	134
Nectarines	−1.6	29	83	0.90	0.49	119
Oranges	−2.2	28	87	0.90	0.46	124
Peaches	−1.6	29	87	0.90	0.46	124
Pears	−2.2	28	83	0.86	0.45	118
Persimmons	−2.2	28	78	0.84	0.43	112
Pineapples	−1.6	29	85	0.88	0.45	122
Plums	−2.2	28	86	0.88	0.45	123
Pomegranates	−2.2	28	77	0.87	0.48	112
Prunes	−2.2	28	86	0.88	0.45	123
Quinces	−2.2	28	85	0.88	0.45	122
Raspberies	−1.1	30	82	0.85	0.45	122
Strawberries	−1.1	30	90	0.92	0.47	129
Tangerines	−2.2	28	87	0.93	0.51	126
Watermelons	−1.6	29	92	0.97	0.48	132
			Vegetables			
Artichokes	−1.6	29	83.7	0.87	0.45	120
Asparagus	−1.1	30	93.0	0.94	0.48	134
Beans (string)	−1.1	30	88.9	0.91	0.47	128
Beans (lima)	−1.1	30	66.5	0.73	0.40	94
Broccoli	−1.6	29	89.9	0.92	0.47	130
Carrots	−1.6	29	88.2	0.86	0.45	126
Cauliflower	−1.1	30	91.7	0.93	0.47	132
Celery	−1.1	30	93.7	0.95	0.48	135
Corn (green)	−1.6	29	75.5	0.80	0.43	108
Cucumbers	−1.1	30	96.1	0.97	0.49	137
Mushrooms	−1.1	30	91.1	0.93	0.47	130
Olives	−2.2	28	75.2	0.80	0.42	108
Onions	−1.1	30	87.5	0.91	0.46	124
Peas	−1.1	30	74.3	0.79	0.42	106
Peppers	−1.1	30	92.4	0.94	0.47	132
Potatoes (white)	−1.6	29	77.8	0.82	0.43	111
Potatoes (sweet)	−2.2	28	68.5	0.75	0.40	97
Tomatoes	−1.1	30	94.1	0.95	0.48	134
			Meat			
Bacon	−2.2	28	20	0.50	0.30	29
Beef (fresh lean)	−1.6	29	68	0.77	0.40	100
Beef (fresh fat)	−2.2	28	65	0.60	0.35	79
Hams and loins	−2.7	27	60	0.68	0.38	86

Table 2 (continued)
SPECIFIC HEAT VALUES OF COMMON FOODS

Product	Average freezing point °C	Average freezing point °F	Water %	Specific heat (Btu/lb °F) Above freezing	Specific heat (Btu/lb °F) Below freezing	Latent heat of fusion (Btu/lb)
Lamb	−1.6	29	58	0.67	0.30	83
Liver	−1.6	29	65	0.72	0.40	93
Pork (fresh)	−2.2	28	60	0.68	0.38	86
Poultry (fresh)	−2.7	27	74	0.79	0.37	106
Sausage (fresh)	−3.3	26	65	0.89	0.56	93
Veal	−1.6	29	63	0.71	0.39	91
			Fish			
Clams	−2.2	28	87	0.90	0.46	130
Cod	−2.2	28	78	0.90	0.49	119
Haddock	−2.2	28	78	0.82	0.43	112
Halibut	−2.2	28	75	0.80	0.43	108
Herring	−2.2	28	70	0.76	0.41	101
Lobster	−2.2	28	79	0.83	0.44	114
Mackerel	−2.2	28	57	0.66	0.37	82
Ocean perch	−2.2	28	80	0.84	0.44	115
Oysters	−2.2	28	87	0.90	0.46	125
Salmon	−2.2	28	64	0.71	0.39	91
Scallops	−2.2	28	80	0.84	0.44	115
Shrimp	−2.2	28	83	0.86	0.45	120
Tuna	−2.2	28	70	0.76	0.41	101

The determination of freezer size which is capable of extracting 3000 Btu/hr/ft would be made by using the following steps:

1. Determine the heat to be removed from one pound of meat. Refer to Figure 10 to determine 60% moisture content.

at −17.78°C (0°F)	116.5 Btu/lb
at 4.4°C (40°F)	30.0 Btu/lb
	86.5 Btu/lb difference

2. Determine the heat load required for the freezing tunnel: 2,700 lb/hr × 86.5 Btu/lb = 233,500 Btu/hr.
3. Determine the belt loading factor (BLF). Since there is a wide variation in belt loading, the effective heat extraction rate of a freezer is established by using the belt loading factor (BLF). This factor is determined by dividing the area of the product distributed on the belt area available. The belt area is established by using the actual belt width on which the product can be placed upon. A simplified formula for calculating the BLF for circular products distributed across a conveyor belt is as follows:

$$\mathrm{BLF} = \frac{0.785 \times (\text{number of patties across belt}) \times (\text{Patty diameter})}{\text{width of conveyor belt}} \tag{4}$$

$$\text{BLF} = \frac{0.785\ (6)\ (5.5\ \text{in.})}{38\ \text{in.}} = 0.68 \tag{5}$$

4. Determine the belt area and tunnel length:* freezer ability × BLF/heat load = belt area; 3000 Btu/hr/ft² × 0.68/233,550 Btu/hr = 114 ft².

$$\text{Approximate tunnel length} = \frac{\text{belt area}}{\text{belt width}} = \frac{114\ \text{ft}^2}{3.16\ \text{ft}} = 36.07\ \text{ft}$$

5. Determine the nitrogen consumption. Since the incoming product temperature is 4.4°C (40°F) and it is important to have a differential of 80° between the inlet temperature and the exhaust gas temperature, the heat absorption capability of 1 lb of nitrogen would be used up to −40°C (−40°F). By referring to the enthalpy chart for nitrogen, the refrigeration value for this condition would be 151 Btu/lb. Therefore, the total amount of nitrogen required per day would be

$$\frac{\text{heat removal/lb meat} \times \text{lb/hr} \times \text{hr/day}}{\text{available refrigeration from } LN_2}$$

$$\frac{86.5\ \text{Btu/lb} \times 2700\ \text{lb/hr} \times 10\ \text{hr/day}}{151\ \text{Btu/lb}} = 15{,}467\ \text{lb/day of } LN_2$$

Or determine the amount of nitrogen required per pound of meat frozen.

$$\frac{\text{nitrogen consumption lb/day}}{\text{production rate lb/day}}$$

$$\frac{15{,}476\ \text{lb/day}}{27{,}000\ \text{lb/day}} = 0.572\ \text{lb } LN_2/\text{lb meat}$$

III. CRYOGENIC GRINDING OF FOODS

Almost from the beginning of man's existence, food was reduced in size to make it more manageable to consume. Wheat was ground to produce flour and hence the size reduction process was firmly established. In our endeavor to improve on the quality of edible products, we are continuously striving to produce a smaller size particle and still retain the original flavor. In normal size-reduction processes, heat is generated when energy is used to fracture a particle into a smaller size. This generated heat usually is detrimental to the product and results in some loss of flavor and quality. When some type of cooling is used to compensate for the heat produced in the grinding process, the food has the ability to remain closely in its original quality. The use of circulating air, cooling water, etc. has to some degree reduced the amount of heat developed in standard grinding mills. Recently, the adaptation of a cryogen such as liquid nitrogen at −195.8°C (−320.4°F) to the grinding process has provided another method of retaining the quality of the food being processed. This extremely low temperature liquid provides the refrigeration necessary to compensate for any generated heat in the mill.

* Select an available tunnel which is commercially available and slightly longer than the calculated estimate.

The following sections describe some of the recent activity in the cryogenic grinding and processing areas.

A. Particle Size Reduction

The regulated science of size reduction of particles, together with a predictable and controllable end result, is referred to as ''comminution''. Comminution is a universally applied process of controlled size reduction to retain such characteristics as taste, flowability, color, and viscosity. The method used to reduce the size of a product is directly related to the hardness, sensitivity to temperture, fracture ability, moisture content, and final particle size desired, etc.

Many products such as foods, spices, chemicals, plastics, and resins which are presently being produced are not adequately reduced in size in general grinding operations. However, low-temperature refrigeration can assist in attaining a finer particle size. Materials which are extremely heat-sensitive can now be processed with the use of liquid nitrogen in a grinding operation.

B. Basic Process

In any size-reduction operation, heat is generated in the mill chamber from the energy of the hammers in their effort to fracture a particle to a smaller size. At times, this generated heat may be sufficient to prohibit the reduction in size of the material. However, with grinding materials which are heat-sensitive, it becomes advantageous to reduce this generated heat by introducing a refrigerant into the mill. Liquid nitrogen which can be sprayed directly into the mill chamber will offset and provide a ''heat sink'' for the heat that is generated during reduction. Occasionally, in order to bring the material to its fracture point, it is necessary to precool the material prior to entry into the mill. This operation can be done in a cryogenic precool conveyor or liquid nitrogen immersion bath system. Figures 12 and 13 illustrate each system. However, refrigeration in the mill may still be necessary to avoid heat build-up and maintain the material at its embrittlement point.

C. Advantages of Cryogenic Grinding vs. Standard Grinding

The use of liquid nitrogen for cryogenic grinding offers many advantages when compared to standard grinding methods that do not use any low-temperature refrigerant. Some of the more common benefits are listed in Figure 14 and are described as follows:

Improved size reduction — With a cryogenic refrigerant, it is now possible to grind certain materials into very fine particles which previously could not be accomplished by other methods.

Reduced particle size — In food products, this directly affects and improves flavor with uniform dispersion when mixed, thereby maintaining the natural quality of the food product.

Fine powders — These improve hopper feeding, color matching in paint pigments, improved flow for packaging, tableting, weighing, etc.

Increased production rates — Precise cryogenic temperature control allows more product to be reduced in size per unit of time.

Reduced heat build-up — With the use of liquid nitrogen there is no product damage due to chemical or property changes. Material softening that leads to mill-clogging is eliminated. Color and flavor are maintained by preventing heat build-up during the grinding cycle.

Higher production and reduced operating costs — Use of liquid nitrogen reduces the mill's electrical power needs, which in turn leads to production increases and reduced operating costs.

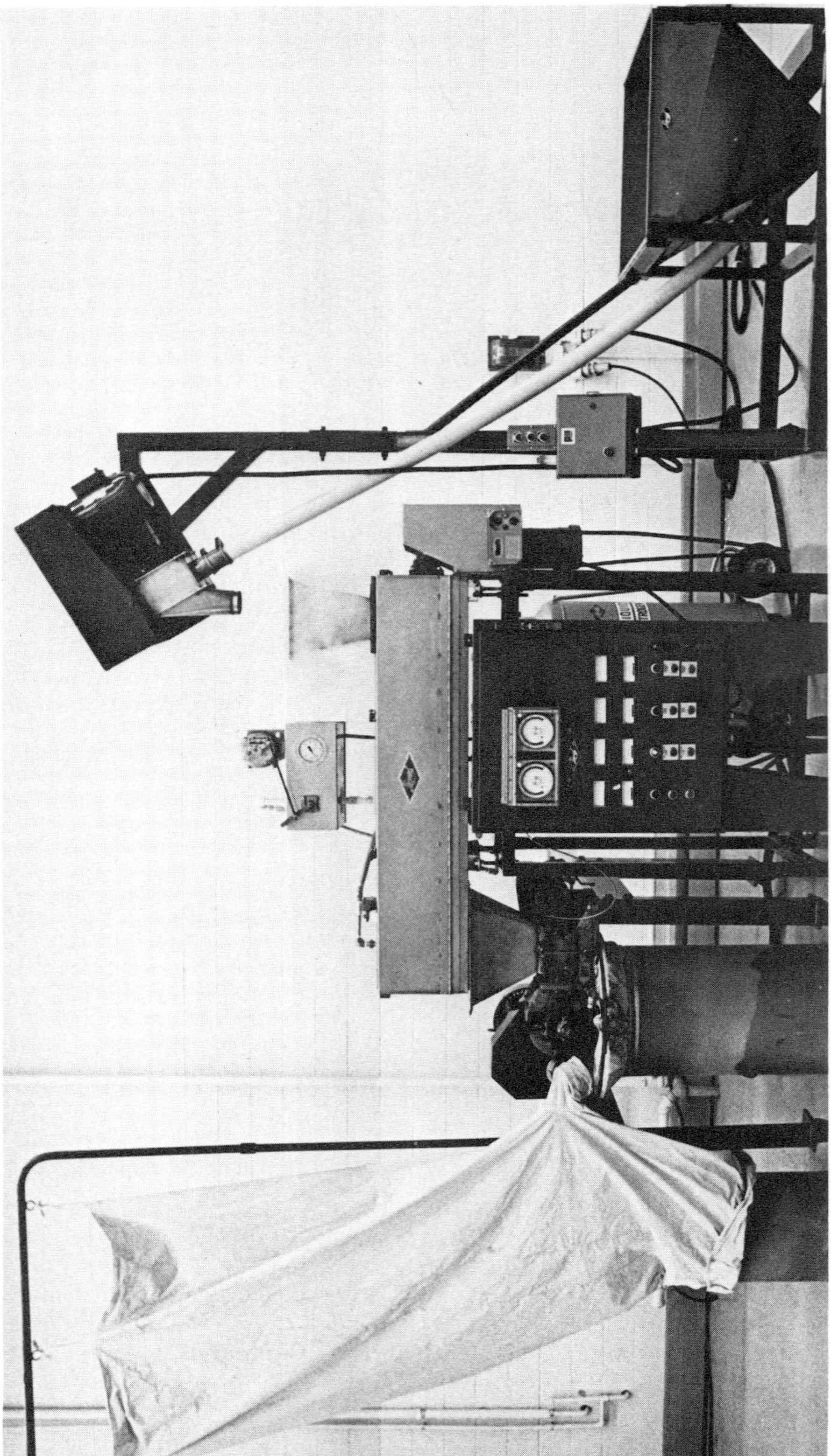

FIGURE 12. Cryogenic precool conveyor system. (Courtesy of Liquid Carbonic Corporation, Chicago.)

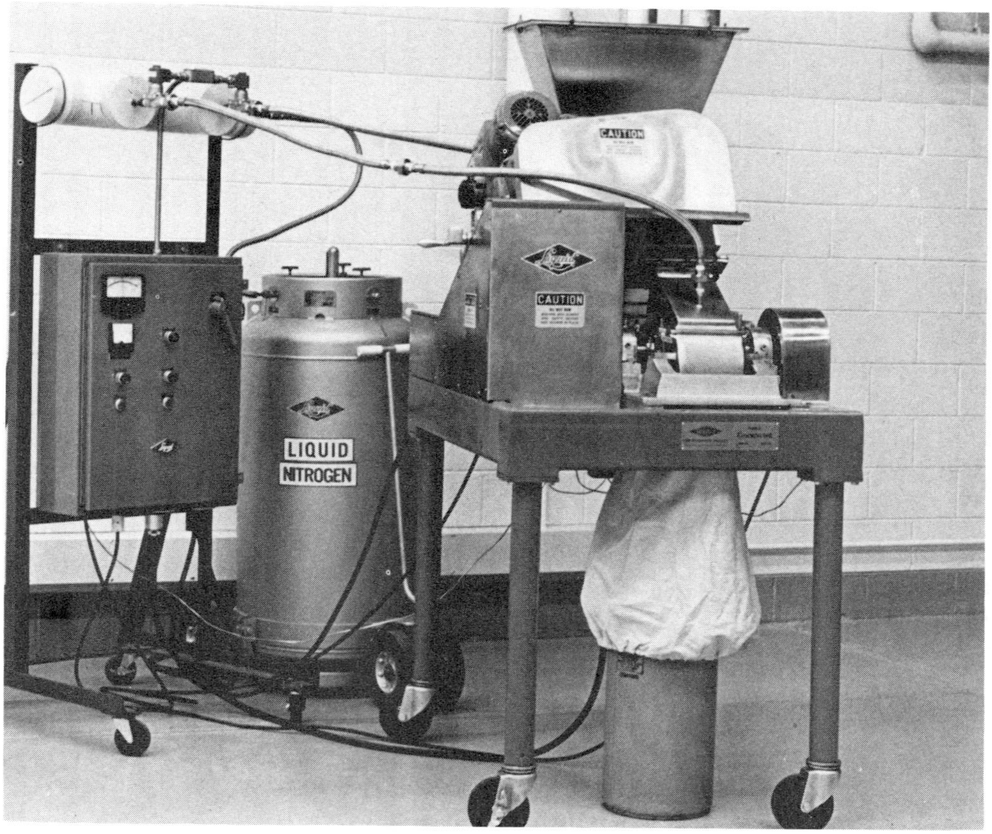

FIGURE 13. Liquid nitrogen immersion bath system. (Courtesy of Liquid Carbonic Corporation, Chicago.)

Practical recycling — Cryogenic grinding allows most previously scrapped or rejected products to be recycled, thereby reducing overall processing costs.

Inert atmosphere — Many materials require a moisture- or air-free atmosphere when being ground. Cryogenic grinding can provide this atmosphere while still providing the refrigeration required. A nitrogen atmosphere can also be used for final moisture-free packaging, which is highly desirable in the food industry as well as others.

D. Spice Grinding

Cryogenic grinding of spices is an acceptable technique that helps to retain the natural quality of spices. The liquid nitrogen which is at −195.8°C (−320.4°F) provides the refrigeration needed to reduce the temperature of the spices and prevent the "volatile oils" which carry the spice flavors from being driven off during the grinding process. It is this retention of volatile oils that preserves the flavor strength per unit of spice close to its natural state. The amount of volatile oils retained by cryogenic grinding as compared to conventional grinding can be determined by simple laboratory comparison.

All spices that are to be ground do not require grinding temperatures as low as −195.8°C (−320.4°F). The temperature to be used is determined by the product, final size, color required, etc. For example, chili pepper could be ground at −115°C (−175°F) while nutmeg, which has a very high oil content, could require a temperature

BENEFITS FROM CRYOGENIC GRINDING

1. OFFSET HEAT BUILD-UP DURING AMBIENT GRINDING
2. GRIND MATERIALS WHICH CANNOT BE PRODUCED DURING AMBIENT GRINDING
3. MATERIALS GROUND INTO FINER AND SHARPER PARTICLE
4. IMPROVE PARTICLE SIZE DISTRIBUTION
5. SCRAP RECLAMATION
6. SEPARATION (LAMINATED MATERIALS)
7. LEADS TO INCREASED PRODUCTION RATES
8. PROVIDES INERT ATMOSPHERE

FIGURE 14. Cryogenic grinding benefits. (Courtesy of Liquid Carbonic Corporation, Chicago.)

of approximately −156.67°C (−250°F). In comparison, oregano would only require a mill temperature of approximately −6.67°C (20°F). Some typical food products that could be cryogenically ground are shown in Figure 15.

The amount of liquid nitrogen consumed per pound of spice ground is directly related to the type of equipment used, operating temperature, and particle size required. It is not possible to predict the actual amount of liquid nitrogen to be used because of uncontrollable operations and widely different requirements. However, some information is provided in Table 3, which can be used as a guide in determining anticipated economical advantages.

E. Grinding Equipment

Numerous types of size-reduction mills are available for processing of such things as food stuffs, plastics, and chemicals. The mills are classified into the following categories: impactor or hammer, attrition, air swept mills. Each type of mill will produce somewhat different particle-size characteristics. In all mills, the final particle size produced is directly related to the following:

1. The size and shape of the screen opening through which the particle must travel.

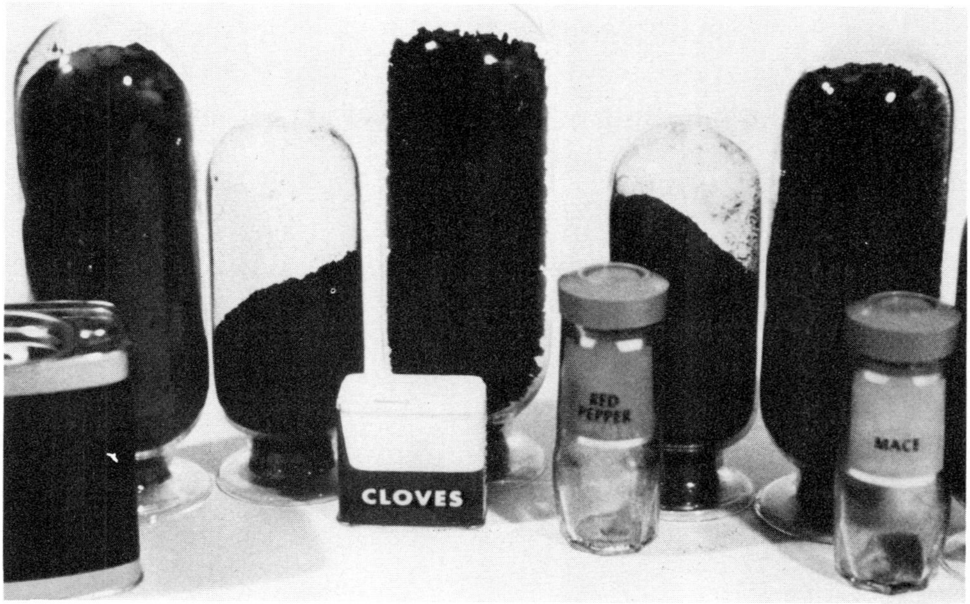

FIGURE 15. Typical food products that could be cryogenically ground. (Courtesy of Liquid Carbonic Corporation, Chicago.)

Table 3
CRYOGENIC SPICE GRINDING DATA

Spice	Mesh achieved %	Mesh achieved Mesh	Production rate (Lb/hr)	Approximate LN_2/lb production
Mustard seed (whole)	100	30	300	0.87
Clove	100	30	300	0.78
Caraway seed (whole)	100	30	300	1.10
Coriander	100	20	150	1.30
Parsley	100	40	110	0.60
Anise seed (whole)	100	30	260	1.05
Basil (French)	100	40	350	0.70
Oregano	100	40	165	0.92
Red pepper	100	40	400	0.53
Cinnamon	100	40	380	0.58
Mustard bran	90	40	300	0.60
Nutmeg (whole)	100	20	750	0.63
Nutmeg (blanched)	100	40	775	0.68
Chili pepper	100	40	550	0.55

NOTE: Values presented are based on liquid nitrogen immersion techniques. Other operating procedures and equipment will produce different results than shown above. Information presented should be used as a guide only.

2. The speed at which the particle travels. At high speeds, the particle will approach the screen opening at a tangent thus making a round hole appear to be elliptical and creating a smaller particle.
3. The style of blade used will determine the characteristics of the final particle.

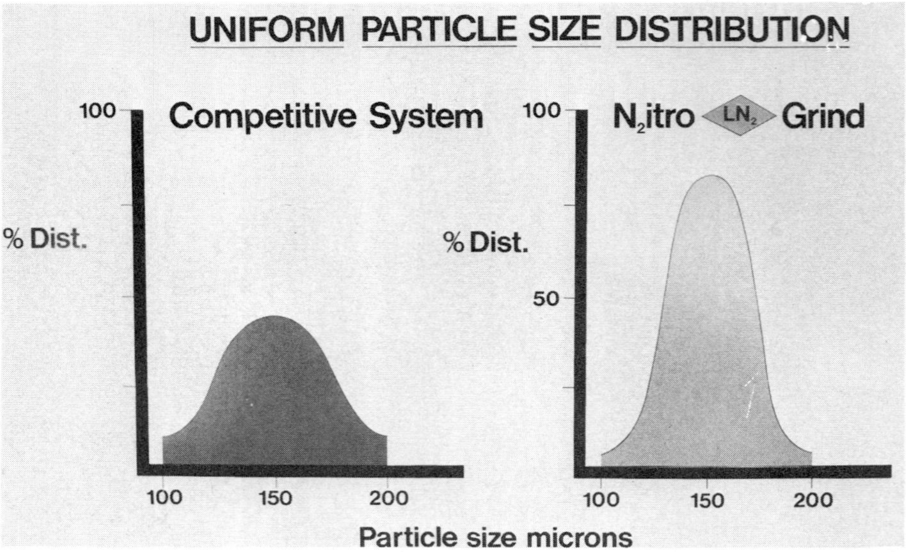

FIGURE 16. Particle size ditribution curve. (Courtesy of Liquid Carbonic Corporation, Chicago.)

For example, flat-edge blades are used for pulverizing, emulsifying, etc., and knife-edge blades are used for granulating, chopping, sizing, and so on.

Typical comparison distribution curves for different types of mills are shown in Figure 16.

Most mills presently used can be adapted to use liquid nitrogen for refrigeration to some degree. Generally, mills that are used for food processing are usually constructed of stainless steel material which is applicable for low temperatures. Other mills fabricated from carbon steel can be considered for modification when a given amount of caution is applied. Carbon steel mills should not be subjected to extremely low temperatures for long durations. However, it is possible to use liquid nitrogen injection into the mill to just overcome the heat being generated by the mill and not produce temperatures lower than the fracture point of carbon steel.

Simple equipment modifications to existing mills are illustrated in Figure 17. A temperature sensor in the mill allows the remotely located temperature controller to open or close the liquid nitrogen flow control valve, thereby injecting liquid nitrogen into the mill and overcoming the heat being generated. As the mill becomes cold and reaches its preset low temperature, liquid nitrogen stops flowing. The temperature in the mill is maintained within a present range by the on/off flow of liquid nitrogen.

F. Precoolers

Cryogenic precoolers can be used to increase the production rates of existing mills even after they have been modified for simple liquid nitrogen conversion. The precooler is also an excellent device to be used in conjunction with existing carbon steel mills. A cryogenic precooler can be combined with impact, attrition, or air-swept mills (Figure 18).

The function of the liquid nitrogen precooler is to remove all of the required heat from a particle prior to its entrance into the mill chamber. The temperature of each individual particle must be at its fracture point or lower. This critical particle temperature must be low enough to absorb the heat of impact in the mill and still fracture. For example, if a particle enters the mill chamber at a temperature warmer than its

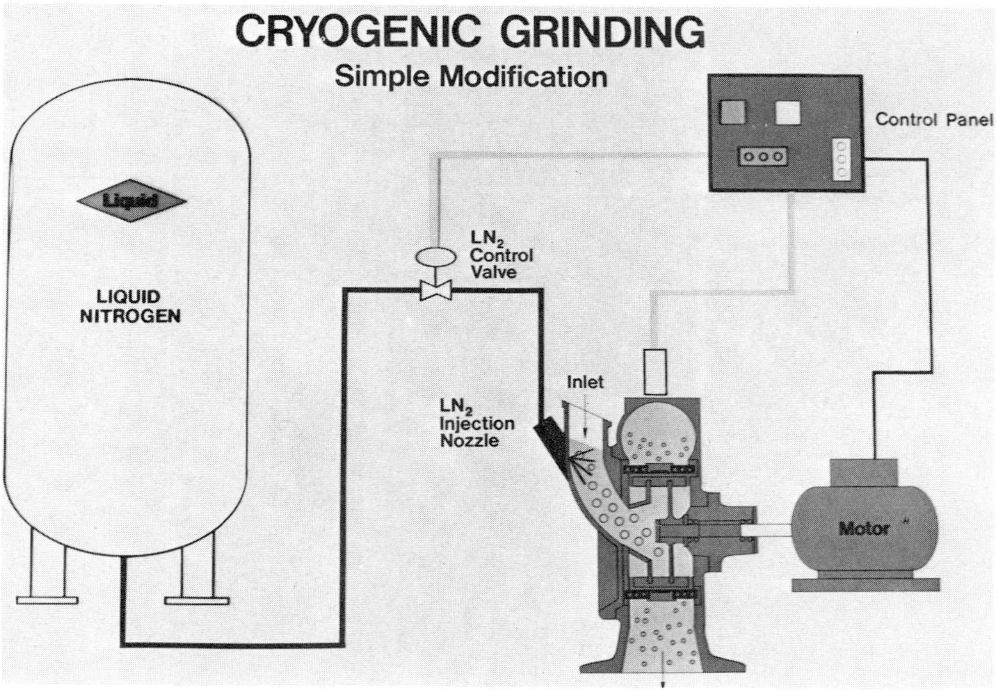

FIGURE 17. Cryogenic grinding simple modification. (Courtesy of Liquid Carbonic Corporation, Chicago.)

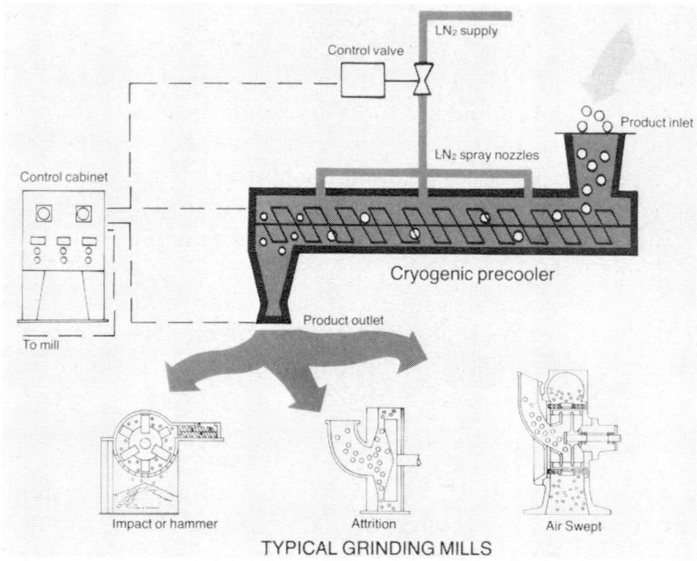

FIGURE 18. Cryogenic precooler and typical grinding mills. (Courtesy of Liquid Carbonic Corporation, Chicago.)

fracture temperature, the heat created by the mill blade upon striking the particle would be transferred to the particle making it rise in temperature and not fracture. As the same particle remains in the mill, it becomes warmer and after repeatedly being struck by the blade eventually melts or smears in the mill.

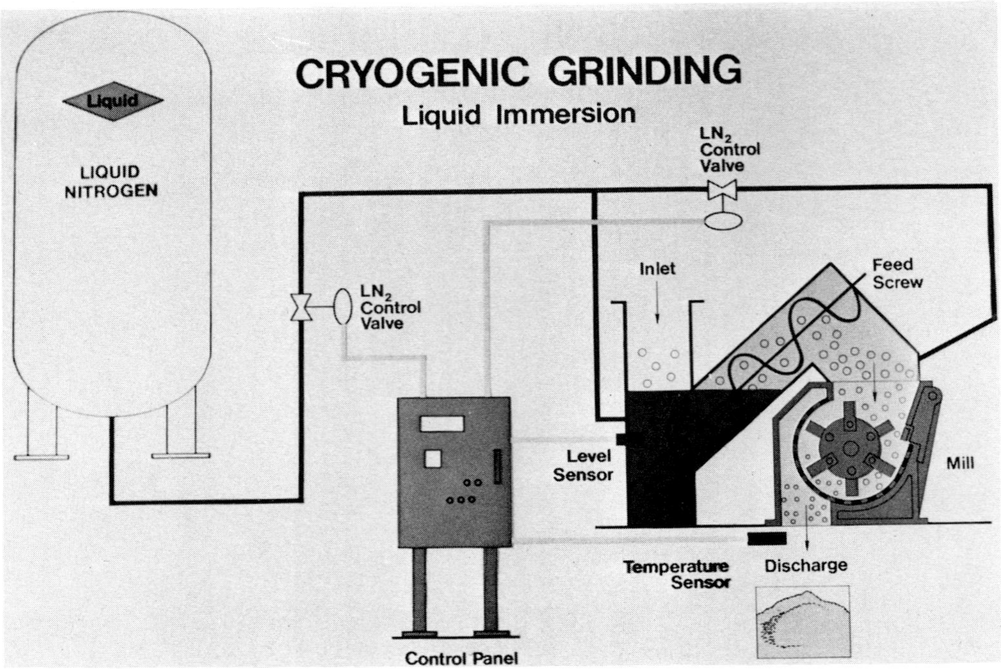

FIGURE 19. Schematic for liquid immersion system. (Courtesy of Liquid Carbonic Corporation, Chicago.)

Cryogenic precoolers therefore must have the ability to reduce the temperature of all of the material before it enters the mill. Also, provisions must be made to control the precooler temperature and feed rate to the mill. Typical flow diagrams of a cryogenic precooler for liquid immersion and a spray system are shown in Figures 19 and 20. Such a system can be adapted to most cryogenic grinding processes which operate on a batch-type principle and moderate production rates. Examples of some materials which can be cryogenically processed and the approximate volumes of liquid nitrogen required are shown in Table 4.

The use of high-volume cryogenic precoolers and grinding equipment is available to process such things as food, chemicals, plastics, and rubber, on a continuous semiautomatic basis. A high-volume production system and a flow diagram are shown in Figures 21 and 22. The continuous large production cryogenic systems are usually designed to utilize the refrigeration available from both the latent as well as the sensible heat of vaporization of the liquid nitrogen. Therefore, high-volume continuous systems generally operate at a relatively high efficiency level.

Examples of other materials which can be reduced in size with the use of liquid nitrogen are shown in Figures 23, 24, and 25. Thermoset and thermoplastic polymers, elastomers, plastic color concentrates, Borox, and other chemicals are shown in Figure 23. Scrap or waste material such as metal, rubber, nylon, polyvinyl chloride, polypropylene, etc. are shown in Figure 24. Cryogenic grinding can also be used in the base materials used in cosmetics, pharmaceuticals, waxes, adhesives, organic dyes, etc. and are shown in Figure 25.

IV. CRYOGENIC BATCH-COOLING OF FOOD PRODUCTS

In the food processing industry the majority of fresh products produced are com-

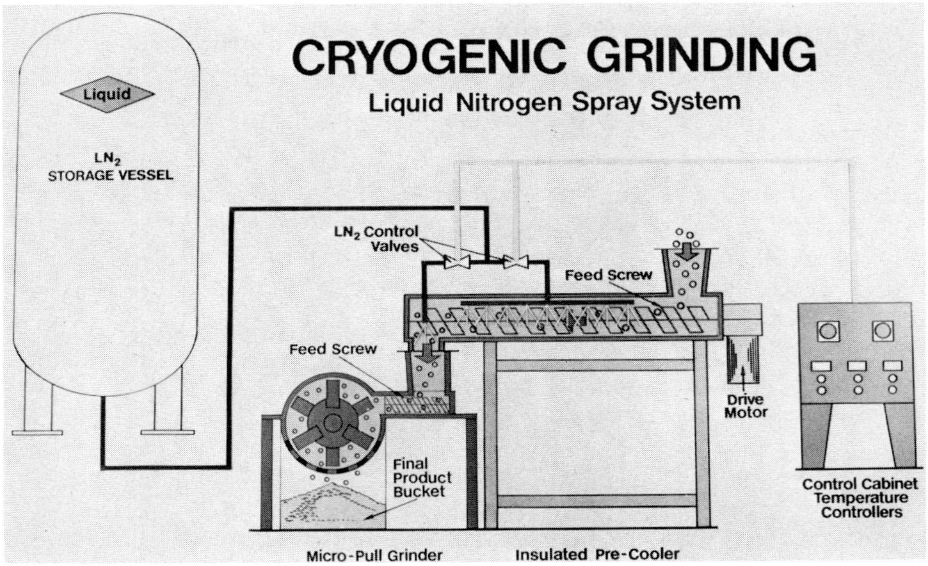

FIGURE 20. Schematic for liquid spray system. (Courtesy of Liquid Carbonic Corporation, Chicago.)

Table 4
EXAMPLES OF CRYOGENIC GRINDING

Material	Liquid nitrogen usage[a] (Lb/lb of product)
Acrylic	0.60 to 0.85
Nylon	1.02 to 1.80
Rubber	0.50 to 0.75
Spices	0.30 to 1.20
PVC	0.50 to 0.70
Polyethylene	0.70 to 0.85
Polyester	0.45 to 0.65
Wax	0.20 to 1.30
Color concentrate	0.25 to 1.20

NOTE: Variation is directly related to individual equipment, desired particle size, etc.

[a] Average values.

prised of numerous items mixed and blended together prior to cooking. Many of the fresh products being considered must be maintained at relatively cold temperatures to maintain their original quality, taste, and so on. It is a well-known fact that within the fresh meat and poultry industry the temperature of the food products must be maintained below a maximum setting. In the batch-mixing of foods such as meat or poultry in which additives such as spices and fillers are added followed by a mixing cycle to achieve a homogeneous mass, the heat generated in the blending, mixing, or grinding equipment can cause considerable loss of quality (Figure 26).

There have been many techniques developed to offset the temperature increase of the batch. Some operations will depend upon water jacket sidewalls on the mixers,

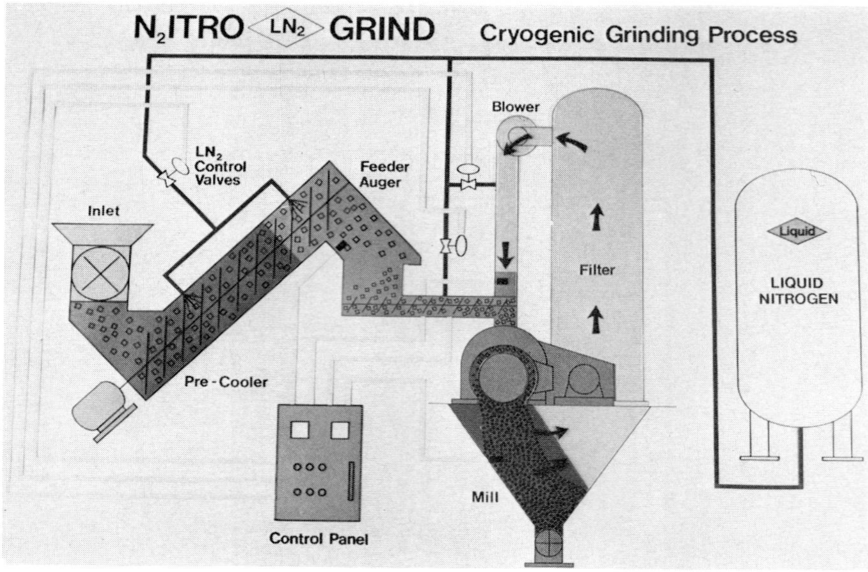

FIGURE 21. Schematic for high-volume production system. (Courtesy of Liquid Carbonic Corporation, Chicago.)

and others rely on colder temperatures of the incoming product. With each of the mentioned techniques, there is very little control of the batch temperatures, and the texture of the final meat and poultry product will be a function of the batch temperature. Should the batch temperature become too warm, a "smear" condition will be formed; similarly, if the batch is too cold, the additives will not mix properly, and it will be difficult to obtain a homogeneous mix.

The use of a cryogen such as liquid nitrogen can be used as a simple, extremely cold refrigerant to control batch temperatures. Since liquid nitrogen is nontoxic and inert, it can be used in direct contact with the food products in the mixer. With the appropriate selection of liquid nitrogen-dispensing spray nozzles, manifold header, and temperature or time controls, it is possible to maintain the desired temperature of the meat or poultry while in the blender, mixer, or grinder. A commercially available blender which has been modified for liquid nitrogen temperature control is shown in Figure 27. The blender is equipped with a hood which is closed during the cryogenic dispensing cycle and is further equipped with an exhaust attachment to effectively and easily remove vapors from the mixer.

The economics of cryogenic batch-cooling can be determined by using the following equation:

$$LN_2 = W(\text{lb}) \times Cp\ (\text{Btu/lb/°F}) \times \Delta T\ (\text{°F})/H\ (\text{Btu/lb})$$

LN_2 = the weight of liquid nitrogen required in pounds to cool the mass of product
W = weight in pounds of the mass to be cooled
Cp = specific heat of the product (Btu/lb/°F)
ΔT = temperature change of the product in °F
H = enthalpy or heat removal capacity of the liquid nitrogen

It is necessary to know the specific heat of the product which is to be cooled and the prevailing cost of liquid nitrogen. The specific heat values will be different for each product. They are directly related to fat, water, spice content, etc. For simplicity, fresh (fat) beef with 65% water content would have a specific heat value of approximately

FIGURE 22. High-volume cryogenic grinding system. (Courtesy of Liquid Carbonic Corporation, Chicago.)

0.60 Btu/lb/°F above freezing. Poultry with a 74% water content would have a specific heat value of approximately 0.79 Btu/lb/°F above freezing.

V. NITROGEN USE IN FREEZE-DRYING

It has been said that the existence and advancement of man is dependent on his ability to preserve his food from day to day and season to season. By applying his increasing scientific knowledge to new and better methods of preservation, man has been able to advance rapidly in this field during the past few years.

The most obvious and widespread forms of food deterioration are due to microbial attack. Bacteria, yeasts, and fungi will compete with man for his food if permitted.

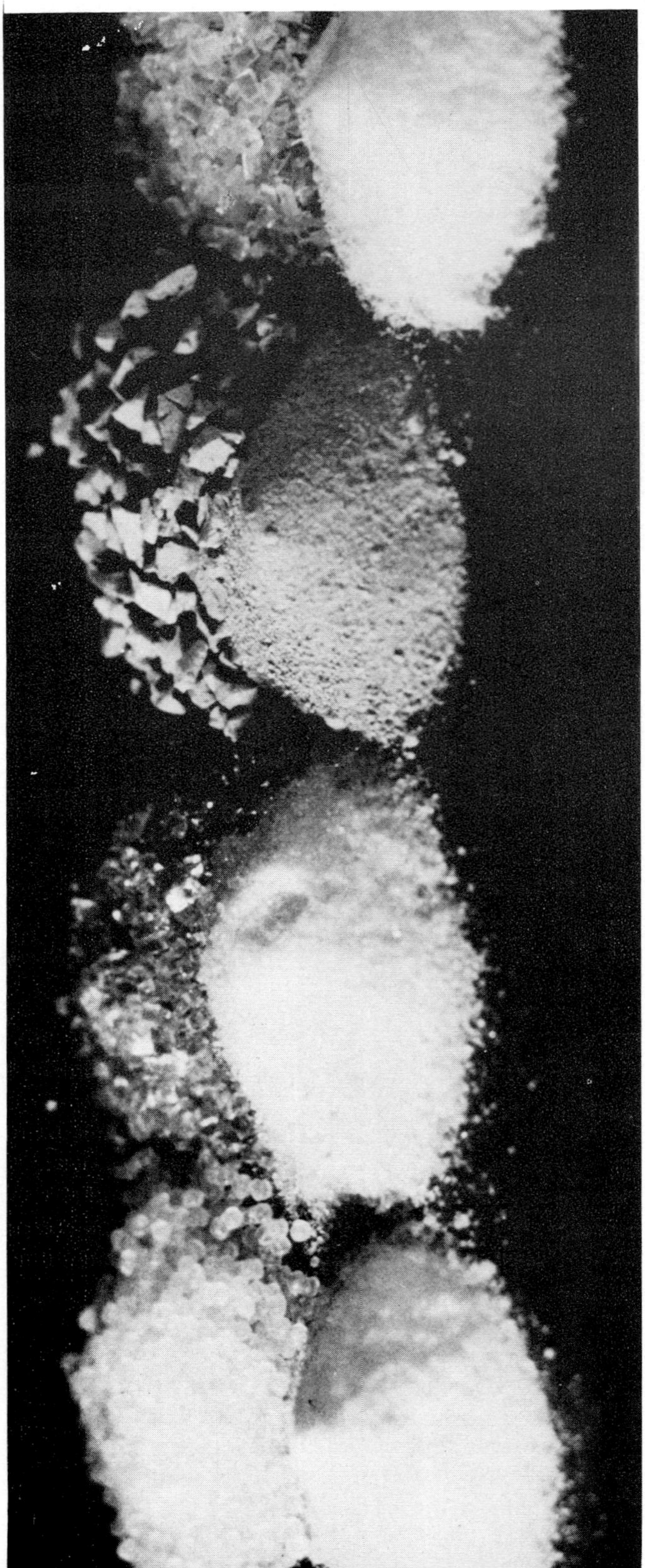

Figure 23. Chemicals shown before and after processing with liquid nitrogen. From left to right, polypropylene, polyvinylchloride, rubber, and nylon. (Courtesy of Liquid Carbonic Corporation, Chicago)

FIGURE 24. Scrap materials processed with liquid nitrogen. Figure shows insulation material, polyethylene, and polyvinylchloride separated into components. (Courtesy of Liquid Carbonic Corporation, Chicago.)

FIGURE 25. Cosmetics and other products processed with liquid nitrogen. (Courtesy of Liquid Carbonic Corporation, Chicago.)

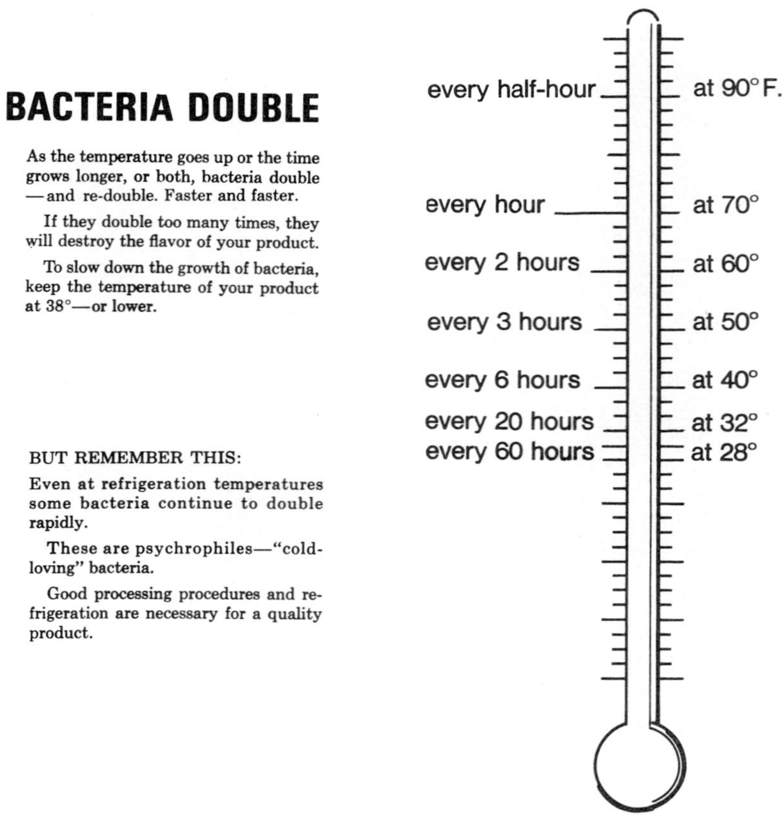

FIGURE 26. Bacteria doubles as temperature increases. (Courtesy of the Poultry and Egg Institute of America.)

Man, sometimes without knowledge of what he was doing, has been dealing with this problem since the introduction of the most elementary forms of food preservation.

Perhaps the oldest method of preservation is drying or dehydration. In more recent times, freeze-drying has become recognized as one of the most attractive methods of dehydrating food since it has the least effect on the flavor, structure, nutritive value, appearance, and color of the final product.

Freeze-dried foods, when properly packaged, are light in weight. They have a long storage life and can be transported and stored without refrigeration. In contrast with other dehydrated foods, freeze-dried products require only minutes to reconstitute to a condition closely resembling the fresh product.

A. Freeze-Drying

Sublimation is the technical term for the familiar phenomenon by which moisture is removed from a frozen material without passing through the liquid state. Freeze-drying is the processing application which uses sublimation for drying under carefully controlled conditions.

The process is essentially quite simple. Materials to be dried are first frozen solidly and are then subjected to a high vacuum and controlled heat input to the product. Under these conditions, water, in the form of vapor, is removed from the product without passing through an intermediary liquid phase (sublimation). Solid components remain locked into position, undisturbed during the drying. The results are a dried product usually indistinguishable in biological activity, size, color, flavor, and shape from the original material when reconstituted.

FIGURE 27. Commercially available cryogenic blender. (Courtesy of Liquid Carbonic Corporation, Chicago.)

Those factors affecting the finished product in packaged form and associated with the deterioration of dehydrated foods which can be influenced by packaging are: moisture uptake, oxygen uptake, flavor contamination, and mechanical damage. Moisture uptake and oxygen uptake are discussed in the following paragraphs.

1. Moisture Uptake

In the freeze-drying process, the final product is a rigid porous material, having a considerable surface area exposed to the surrounding atmosphere. The moisture content generally lies in the region of 2 to 4%, which corresponds to a very low relative humidity. Such products are extremely hydroscopic and in contact with an environment of greater relative humidity, will absorb moisture rapidly. Consequently they must be isolated, as efficiently as possible, from the surrounding atmosphere and packaged in a moisture-free atmosphere.

The maintenance of this dry state is of twofold importance. First, the major deteriorative chemical reactions in dehydrated foods (notably nonenzymatic browning) are retarded at very low moisture levels (oxidation is an important exception). Any increase in moisture content will increase the rate of these reactions, thus shortening the shelf life of the product. Secondly, the fundamental function of drying as a method of food preservation is to inhibit the growth of microorganisms. Whether or not a microorganism will grow is governed by the equilibrium relative humidity of the food product. Bacteria, in general, require more than 90% relative humidity and molds more than 70%. Most dehydrated foods have an equilibrium relative humidity far below these limits. Their proper packaging is necessary to maintain these low safe limits.

The relationship between equilibrium relative humidity and moisture content is peculiar to each product, and it is important that this relationship be known. For example, it will allow a calculation to be made relating to the moisture permeability of a packaging material to a safe shelf life. The amount of water absorption that can be tolerated by a product is determined by studying the relationship between its moisture content and its equilibrium relative humidity.

Two methods are available for determining the moisture permeability of films or packages:

1. A piece of the film is used to seal a cup containing a desiccant. The cup is stored in a constant humidity chamber (relative humidity 75%) and weighed periodically. Since the area of the film and water transmission rate are shown, permeability may be calculated.
2. Sample packs made from the film and containing desiccant are sealed and placed in a chamber with relative humidity of 75%. Weight increase with storage time is noted as above.

Method 2 is favored since actual packages may contain creases, folds, and seals which cannot be accounted for in Test 1. They may greatly impair the performance of some films.

2. Oxygen Uptake

Dehydrated foods are susceptible to deterioration when exposed to oxygen. Meats and other products containing fats develop rancidity. There is sufficient lipid material even in potatoes for strong rancid flavors to develop when oxidation of the dehydrated product takes place. Most green vegetables develop off-flavors, frequently of a hay-like nature, and some bleaching of the color may occur. Exceptions among vegetables seem to be peas and French beans, which seem to deteriorate very little when packaged in air for periods up to 1 year. The effect of oxygen on dehydrated fruit indicates they are generally less sensitive than vegetables.

Also associated with the effects of oxygen is the effect of light, which in some cases may step up oxidation and adversely affect the color of both meats and vegetables.

The exact tolerance of each food product should be determined and the minimum level of protective packaging specified. Accelerated storage tests (i.e., for a short time at abnormally high temperatures) are always open to some doubt in interpretation. The best method of discovering the oxygen tolerance of foods is to store them for their intended maximum storage life at different levels of oxygen concentration. Although a taste-panel evaluation has inherent weaknesses, it is, nevertheless, a realistic assessment of quality changes and can indicate the status of acceptability.

Once oxygen tolerance limits have been established, suitable films may be chosen giving the desired storage life. Methods for determining the permeability of films to oxygen are fairly complex and vary greatly. By filling a package with nitrogen, sealing it, placing it in an oxygen atmosphere, and analyzing the contents at intervals, the probable shelf life of the product can be determined.

B. Vacuum-Breaking With Nitrogen

The nature of the freeze-dried product is such that inert packaging with nitrogen may not provide all of the desired protection. Upon completion of the drying operation, breaking of the vacuum with air will result in a product with cells that contain air. Vacuum-breaking with nitrogen, however, will remove all the air surrounding the product and actually displace much of the air originally contained in the cells of the product.

In an attempt to remove all of the air from the cells of the product, many major producers of freeze-dried products are now breaking vacuum in their drying chambers with nitrogen.

C. Summary

Since moisture and oxygen are the two major causes of deterioration in freeze-dried products, it is necessary to eliminate them from the final product. Breaking vacuum with nitrogen and packaging the product in a nitrogen atmosphere eliminates the oxygen and, at the same time, provides a storage atmosphere with an extremely low moisture content −67.7°C (−90°F) dew point or 3.5 ppm water by volume.

REFERENCES

1. **Liquid Carbonic Corporation,** Food Freezing With Liquid Nitrogen, Tech. Sales Bull. No. IA-1017, Liquid Carbonic Corp., Chicago.
2. **Liquid Carbonic Corporation,** Freeze Drying, Tech. Sales Bull. No. IA-1022, Liquid Carbonic Corp., Chicago.
3. **Liquid Carbonic Corporation,** Flash Freezing With Liquid Nitrogen, Form No. 6072-R-75, Liquid Carbonic Corp., Chicago.
4. **Liquid Carbonic Corporation,** N_2itro LN_2 Chill Batch Chilling, Form No. 6624, Liquid Carbonic Corp., Chicago.
5. **Liquid Carbonic Corporation,** Liquid Nitrogen Plastic & Resin Grinding, Tech. Sales Bull. No. IA-1075, Liquid Carbonic Corp., Chicago, 1976.
6. **Liquid Carbonic Corporation,** Cryogenic Food Freezing, Cryogenic vs. Blast Freezing, Form No. 6342, Liquid Carbonic Corp., Chicago, 1972.
7. **Liquid Carbonic Corporation,** N_2itro Grind, Cryogenic Grinding Pre-Cooler, Form No. 6648, Liquid Carbonic Corp., Chicago.
8. **Liquid Carbonic Corporation,** N_2itro Grind, Cryogenic Grinding Systems, Form No. 6625, Liquid Carbonic Corp., Chicago.

Chapter 10

RESEARCH AND DEVELOPMENT STUDIES IN CRYOGENICS

N. Braton

TABLE OF CONTENTS

I. INTRODUCTION

The purpose of this chapter is to provide an overview of current studies in the field of cryogenics. Though some of the material does not technically qualify as a case study, we have nonetheless used the format of a case study to present the information in a brief and concise manner. In several instances, the material is further complimented by elaboration in the appendixes.

The application of cryogenics to recycling and processing is so new that this list of studies is incomplete. The uses listed are those that have been tried. Some have received immediate acceptance while others must wait until the economy or the thinking of the public is more favorable to conservation. These studies are intended to motivate and inform those persons seeking new solutions for the age-old problem of recycling.

Discussion of the future of cryogenics is contained in the conclusion to this chapter.

II. CRYOGENIC PROCESSING OF THE TIRE

A. Case History

1. Problem

The automobile/truck tire has been one of the more difficult items to recycle or — even worse — to discard. This resulted in billions being discarded along the countryside, behind buildings, in the woods, or anywhere.

2. Changes Made Possible Because of Cryogenics

Cryogenics provides the necessary technology for the effective recovery, separation, and reuse of all materials used in the tire. In fact, the use of cryogenics is the only known way to recover the rubber from the steel radial tire (Figure 1).

3. Advantages and Disadvantages

The advantage is an effective and efficient technology for material recovery from used tires. The major disadvantage is the cost of the cryogen. However, as the market for the end products develops, the cost for the cryogens will become less noticeable.

4. Acceptance

The cryogenic processing technique for tires has received international attention. For a more complete report on the subject, see Appendix I.

B. Applications of Rubber Reclaimed Through Cryogenics

1. Manufacture of Containers

a. Problem

Containers such as flower pots are currently being made from a large variety of materials. There appears to be no scarcity for such materials. However, the fact that used rubber and plastics are now being used to mold these units is another use for recyclable materials.

b. Changes Made Possible Because of Cryogenics

Cryogenics aids in making rubber from used tires available for recycling purposes.

c. Advantages and Disadvantages

Studies have not been made to determine whether or not a rubber plant pot is advantageous over conventional pots (Figure 2).

d. Acceptance

Three plants in the world are producing these items.[1]

2. Manufacture of Lawnmower and Edger Blades

a. Problem

One of the more hazardous pieces of equipment around the home is the lawnmower. A 1975 Stanford Research Institute study shows that during the 1973 to 1974 season there were more than 190,000 mower-related accidents in the U.S. Three fourths of these accidents resulted from direct contact with the blade. The others resulted from flying objects thrown by the revolving blade. The accident-related costs for this period exceeded $400 million. The U.S. Public Safety Commission estimated that in 1976, 56,000 people received hospital treatment from mower injuries with estimated costs having been $327 million.

FIGURE 1. The photo shows a volume reduction of 82% for the rubber of a tire when cryogenically processed.

FIGURE 2. Plant pots are manufactured from 100% recycled materials, including rubber from the automobile/truck tire. The above pots are being manufactured in Southlake, Tex.

b. Changes Made Possible Because of Cryogenics

The rubber blade reduces the chances for accidents. When the revolving blade comes in contact with such objects as the toe of a shoe, the blade flexes, which reduces the chance for serious injuries. The same principle applies when the spinning blade contacts stones. The blade bends and passes over the top. The rubber blades are shock-absorbent. The blades are made by molding used rubber around a metal bar and a strip of nylon mesh. The metal is considerably shorter than is the blade and is useful when fastening it to the mower spindle. The nylon extends the full length of the blade

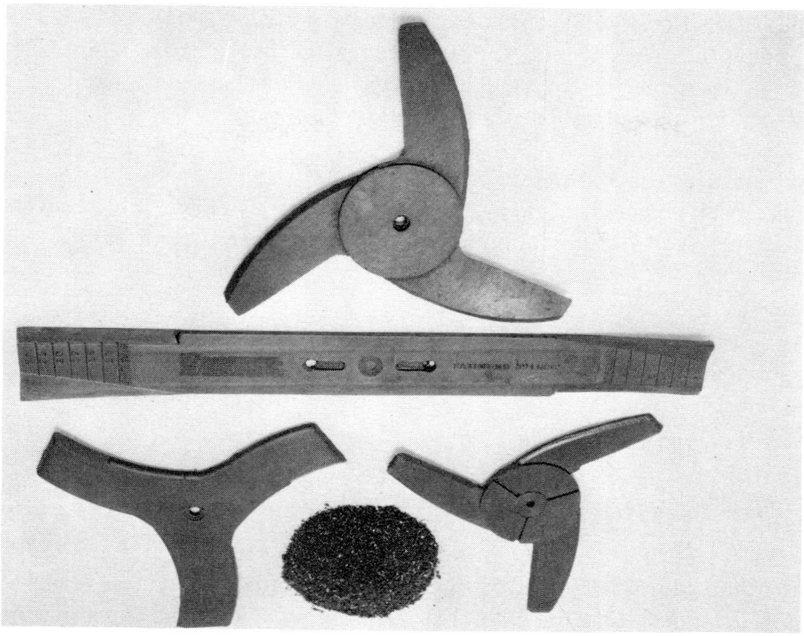

FIGURE 3. The rubber lawn mower and edger blades reduce the chances for injury and provide for greater uses of recyclable materials.

and becomes the self-sharpening cutting edge. The rubber is mixed with an adhesive similar to that used in tire recaps; under pressure and heat the bonding occurs, which is necessary when forming the blade (Figure 3).

c. Advantages and Disadvantages

The advantages and disadvantages of rubberized blades are

1. These blades perform equally or as well as the steel blade. It reduces the accident rate and aids in using of salvageable material.
2. Rubber blades eliminate the "bent crankshaft" problem which is so common when a steel blade comes in contact with large rocks or other stationary objects.
3. Revolving rubber edger blades can be brought into direct contact with the sidewalk or driveway, thereby removing the unwanted grass/weeds from the expansion joints or cracks without generating sparks and flying particles of concrete. If the pavement is asphalt, this material could easily be damaged with a steel blade, whereas it is unaffected with the rubber blade.

d. Acceptance

These blades are patented and being manufactured by the Safe-T-Blade Company of Phoenix, Ariz. and are available on the market. For American National Standard No. B71.1a-1974 "Safety Specifications for Power Lawn Mowers Lawn and Garden Tractors and Lawn Tractors", see Appendix II.

3. Improved Railroad Crossing and Bridge Surfaces

a. Problem

The general public associates railroad crossings with roughness. The maintenance costs for railroad crossings are high and the cycle short.

b. Changes Made Possible Because of Cryogenics

The application of cryogenics for the recovery of used rubber from such hard-to-process items like steel belted tires has made available adequate supplies of rubber materials for improved or rubberized crossings.

c. Advantages and Disadvantages

Rubberized rail crossings reduce the discomfort to passengers and reduces automobile maintenance such as wheel alignments, shock absorbers, and tires.

d. Acceptance

Rail crossings made of rubber are becoming quite common. Several such crossings can be examined at both Madison and Janesville, Wis.[2]

Further information on the use of rubber for resurfacing bridges can be found in Appendix III.[3]

4. Soil Conditioner

a. Problem

Millions of acres of potentially arable land throughout the world are nonproductive because the land cannot efficiently utilize the water which is avaiable either as precipitation or by irrigation. Often the soil has such high clay content that the water cannot be absorbed and runs off.

b. Changes Made Possible Because of Cryogenics

Cryogenics makes rubber available in abundant supply which, when properly processed, can be used to condition the soil for productive crops.

c. Advantages and Disadvantages

Conventional cryogenically processed rubber particles will not stay mixed with the soil so they are not a good soil conditioner. To achieve a soil conditioner, the rubber particles first must be fermented.

The fermented passenger tire mesh (PTM) is completely different from the starting material. Most of the oil has been removed and the product is not hydrophobic; rather, it absorbs water readily. A substantial amount of oxygen has been added; the styrene ring has been opened and the oxygen incorporated in carboxyl groups. The product is a cationic exchange resin; particle size has been reduced to 250 to 300 mesh. This combination of properties makes the fermented product a soil conditioner.

d. Acceptance

For further information, see Appendix IV.[4]

5. Fuel

a. Problem

An estimated 250 million tires are scrapped annually. If these tires were converted to energy, they would provide the Btu equivalent of 3.75×10^6 tons of coal or 6.94×10^8 gal of oil.

b. Changes Made Possible Because of Cryogenics

Now that the tire can be processed cryogenically, rubber can be made available in quantities necessary for continuous supply to the boiler.

c. Advantages and Disadvantages

Used rubber is equal to or better than fuel oils and gases in heat values. The huge backlog of tires can definitely be used to alleviate some of the current fuel shortages. For further information on the subject of using recycled rubber as a fuel, refer to Appendix V.[5]

6. Manufacture of Irrigation Hoses

a. Problem

Irrigation by sprinkling systems is an inefficient method of watering. When the weather is dry and hot, it is reported that 75% of all water passing through a conventional sprinkler system is lost to evaporation.

b. Changes Made Possible Because of Cryogenics

The irrigation hose in Figure 4, made totally from recycled materials (rubber, in particular), dispenses water by the osmosis principle. The hose is buried approximately 3 in. below the ground surface. Using 5 lb of water pressure, moisture is emitted through the hose wall and absorbed into the soil with no loss due to evaporation.

c. Advantages and Disadvantages

This product saves endless quantities of water and improves crop production. Most plants do better when watered at the roots than when sprayed onto the leaves. The product consumes large quantities of salvaged materials.

d. Acceptance

There are reportedly three plants in the world producing this kind of irrigation hose.[2]

7. Oil Spill Cleanup Operations

a. Problem

It is reported that there are thousands of oil spills annually. The costs of oil spills to wild life and nature are enormous. Reports indicate that technology is available for collecting and containing oil spills on the water, providing the water remains relatively calm. However, when this oil can no longer be contained, it often works its way toward the shore. Once the oil gets to shore, major and expensive cleanup operations become necessary. The current cleanup techniques for beach work are antiquated and at best slow, tedious, and quite inefficient.

b. Changes Made Possible Because of Cryogenics

Mixtures of cryogenically processed rubber particles have been tested as absorbent materials for hydrocarbons. Normally, the weight of oil or hydrocarbons absorbed ranges from three to eight times the weight of the absorbent mixture. The amount absorbed depends on the type of hydrocarbon. The rubber particles can be packaged in porous bags which contain the particles. The wall construction of the bag allows hydrocarbons to diffuse and flow into the rubber particle absorbent. Once the bag absorbs the hydrocarbon, little migrates out of the bag. This allows for proper and relatively easy disposal.

The recycled rubber particles were obtained by cryopulverization of tires, although tire tread regrounds have also been tested and found appropriate. Suitable particle sizes range from 20 to 70 mesh. The particles are encased in a porous fabric which is made from nonwoven polypropylene plastic fibers (Figure 5). This fabric is commonly used today in diaper products. The fabric has several attractive properties for this application. It has suitable strength, correct porosity to prevent the unswollen rubber

FIGURE 4. This irrigation hose is manufactured from 100% recycled materials, including rubber from the automobile/truck tire. The above is being manufactured in Southlake, Tex.

FIGURE 5. One of four configurations developed by personnel at the University of Wisconsin, showing ribbed mattress-type design. The bag rubber floats, and the cleavages help collect the oil as the waves wash it into the bag.

particles from migrating through the fabric, low density (less than water so it can float on water surfaces), and good sealing characteristics.

Several uses of these bags are proposed. Gasoline spills at airports or gasoline stations could be readily absorbed and disposed of. Consumers can use these porous bags as absorbents for crankcase and oil leaks from vehicles. The bags could also be used for oil spill cleanup on water surfaces and beaches. Final disposal of the bags could be incineration to recover the energy or destructive distillation to recover the oil. A squeeze pressure process will recover up to 60% of the oil absorbed.

Where the oil has already reached the beach, a different approach must be taken. Once the oil has been absorbed by the sand, it has been found that by spreading cryogenically processed rubber particles over the contaminated area followed by a treatment of liquid nitrogen, the oil-soaked sand becomes a solid and can be lifted away from the subsurface or pure beach sand.

c. Advantages and Disadvantages

Several advantages of bagging the rubber are

1. The process is clean and fast when compared to current practices for oil spill cleanup.
2. The bags can be easily transported and stored indefinitely for emergency use.
3. The cost for materials, processing, and filling is low.
4. Much of the oil can be reclaimed by using pressure or centrifugal force.

d. Acceptance

Interest is rapidly generated by oil companies, environmental groups, Departments of Natural Resources, and other responsible parties.[7]

8. Improved Roofs and Driveways

a. Problem

Most conventional asphalt-type flat roofs develop cracking and leaking problems. The sun tends to dry out the asphalt, and the low temperatures tend to make it brittle. The continual expansion/contraction cycles caused through heating and cooling, coupled with loading from walking to snow loads, causes cracks to take place with subsequent leaking.

b. Changes Made Possible Because of Cryogenics

Because rubber from tires and other waste products can now be recovered quickly and efficiently using cryogenic techniques, the supply is ample for such items as built-up flat roofs. As the supply of used rubber becomes more abundant, so does the technology for its use.

c. Advantages and Disadvantages

Rubberized roofs sharply reduce maintenance costs. A flat roof containing rubberized asphalt could have a maintenance-free cycle of 10 years as compared to 2 years, which is normal with conventional asphalt roofing materials. The optimum level of rubber-to-bitumen blend is somewhere between 5 to 25% mix.

With a 10% rubber-to-bitumen mix, each gallon of emulsion would contain approximately 14 g (½ lb) of ground rubber. One gal of emulsion would normally cover 12 m^2 (40 ft^2). Each used passenger tire produces on the average 45 N (10 lb) of rubber or enough to prepare 75 ℓ (20 gal) of emulsion. If the average driveway is 2.4 m × 30 m (8 ft × 100 ft), it would take 75 ℓ (20 gal) of emulsion and use the rubber from one tire.*

* Scrap rubber tire utilization in road dressings. EPA Technology Service — EPA 670/2-74-014, March, 1974.

9. Mercury Removal From Water

a. Problem

Mercury, one of several classed as heavy metals, is poisonous and of concern to most environmentalists. Mercury is used routinely in industry as a catalyst in nitric acid solution to dissolve aluminum alloy fuels. It is important that all mercury be removed from streams and other bodies of water.

b. Changes Made Possible Because of Cryogenics

Research at the DuPont laboratories has been successful in developing a simple method for removing mercury from aqueous process streams. The system consists of a column packed with finely ground vulcanized rubber preceded by a liquid cyclone separator. In one test, a pilot-scale packed bed containing finely ground tire rubber produced an effluent having < 10 ppb mercury for 170-bed volumes throughout, and < 100 ppb for over 600-bed volumes. Feed solution containing 1800 to 11,000 ppb mercury in a dissolved or colloidal state. In batch laboratory tests, mercury was reduced from 750 ppb to < 5 ppb in 5 min at 70°C.[8]

The application of cryogenics to recycling of used tires makes available adequate quantities of unvulcanized rubber for filtering purposes.

10. Improved Roadways

a. Problem

Traditional asphalt roads have a long history of spring breakup, pot holes, short life, and high maintenance costs.

b. Changes Made Possible Because of Cryogenics

Cryogenic technology makes possible the reclaiming of rubber from tires in general and, more specifically, steel radial tires. This in turn makes rubber available for use in highways, streets, runways, and flat roofs.

c. Advantages and Disadvantages

Tests made by personnel at Battelle Institute, Arizona Highway Department, and others show that rubberized surfaces are far superior in most aspects than its counterpart, the traditional asphalt.

d. Acceptance

The acceptance by governmental agencies to include rubber particles in asphalt mixes for highways, streets, runways, and flat roofs has been most favorable.

Figure 6 shows a comparison of conventional asphalt vs. rubberized asphalt after 3 years of exposure to the elements.

For additional information on the use of rubber in asphalt, see Appendix VI.[9]

11. Improved Animal Mattresses

a. Problem

Cattle and horses housed on concrete floors covered by straw or wood shavings are subjected to bruises and abrasions caused from slipping when getting up and down and when giving birth to their young.

b. Changes Made Possible Because of Cryogenics

Versatile mats are manufactured using a controlled heat and pressure process. Ground rubber is combined with specially formulated additives and vulcanized at high pressure in a modern plant. This unique process gives a textured nonskid surface de-

FIGURE 6. This photo compares conventional asphalt (lower third) to rubberized asphalt (upper third). Note the difference of cracking and no cracking. Courtesy of Charles McDonald, Phoenix, Ariz.

signed to provide outstanding traction and animal comfort properties. Mats furnished the footing necessary to help make rising as simple as possible, thereby helping to reduce rubber burns and bruises.

Thick (¾ in.) mats provide superior insulation which assures a warm comfortable bed between the cows' udder and concrete platform. Cushion comfort provides knee and leg protection which helps prolong the life of the cow and her milk production.

c. Advantages and Disadvantages[6]

The advantages of rubber mattresses over conventional floors are that they

1. Provide greater cow comfort — extendes longevity and production
2. Save 80% or more on bedding use
3. Save time and labor used in conventional bedding, handling, and storage
4. Reduce cow injuries — nonskid surface helps provide sure footing
5. Are sanitary — flat nonribbed surface is easy to clean
6. Are easily installed in existing stalls or new construction
7. Are economical — exclusive manufacturing process using ground rubber provides high quality at affordable prices

d. Economics

Based on a 40-cow herd, a comparison of bale-bedding cost vs. rubber-mat cost can be made. One important point to be made is that the rubber mats can be cleaned and continuously reused.

Bale	Cost
One bale of straw (@ $1 per bale) per cow per week = 2080 bales per year =	$2080
Rubber mat	
One rubber mat per cow per full year =	1720
Cost Difference =	$360

The gross difference of $360 is further increased the second year by reuse of the rubber mats, thus eliminating cost of bedding entirely. Here we find a 100% return on investment plus a profit of $360 the first year of use.

The price of baled straw or wood shavings can be adjusted up or down, depending on its value in your area.

Dairymen, spot-checked throughout Wisconsin, quoted $1 per bale of straw. In Minnesota, wood shavings were quoted as high as $2 per bale. These were average-size bales.*

III. RUBBER DEFLASHING

A. Case History

1. Problem

Thin flash is formed on rubber parts during molding. This flash must be removed prior to use.

2. Changes Made Possible Because of Cryogenics

The thin sections are rapidly chilled and temporarily embrittled while the thick sections remain tough, enabling the thin sections to be broken off by tumbling the parts.

3. Advantages and Disadvantages

This decreases labor requirements but may induce surface cracks.

4. Acceptance

It is widely used.

IV. PYROLYSIS OF CRYOGENICALLY PROCESSED TIRES

A. Case History

1. Problem

Each automobile tire produced uses approximately 8 gal of oil; 5 gal are used in the material, and 3 gal are used as energy in manufacturing the tire. Multiply this by the 240 million tires manufactured annually and we have nearly two billion gal of oil used in the manufacturing of tires in the U.S. The recovered oil from the discarded tire would be useful.

2. Changes Made Possible Because of Cryogenics

The steel used in tires interferes with the process of pyrolysis. The application of cryogenics makes possible the separation of rubber from the steel prior to the pyrolysis process.

3. Advantages and Disadvantages

From each ton of tires can be produced by pyrolysis 3 to 3½ bbl of oil, 575 to 700 lb of carbon black, 85 to 100 lb of steel, and 70 to 80 lb of fiberglass. By cryogenically processing the tire, the steel and fabric can be removed prior to the pyrolysis process. This should speed up the process, produce less waste, and provide a better grade of steel and fabric for other uses.[10]

* Humane Equipment Corp., Baraboo, Wis.

4. Acceptance

The pyrolysis of whole tires or tire sections is receiving a great deal of attention, with several pilot plants in operation. The use of cryogenically processed rubber for pyrolysis has been tried, with favorable results being reported. Before large-scale runs can be made using cryogenically processed rubber, the source of supply must be improved.

V. RECYCLING OF THIN FILM PLASTIC WASTES

A. Case History

1. Problem

Many commercially produced thermoplastic polymers are difficult to grind at ambient temperatures. These plastics are tough, and throughout the process the material heats up. This higher temperature causes the material to reagglomerate, which results in a tackiness and an adherence to the cutting blades. This impairs and slows the cutting process.

2. Changes Made Possible Because of Cryogenics

By reducing the temperature of these thin film materials, their properties change from tough and tacky to hard and brittle.[11]

3. Advantages and Disadvantages

A brittle thin film polymer can be easily and quickly shattered into reduced particle sizes when subjected to the impacts of swinging hammers (Figure 7). The disadvantages may be the additional costs involved for reducing the temperature. However, the process of particlizing thin film plastics is so much improved at reduced temperature than at ambient temperature that the economics can be overcome.

4. Acceptance

The process of using cryogenics in recycling thin film waste polymers is progressing rapidly, and the process is being used in several industries. It is a well-established technology in the area of heat-sensitive materials.

VI. RECLAIMING NONFERROUS METALS

A. Case History

1. Problem

Many items, such as the automobile alternator, generator, electric motor, and solenoid switches are made from steel, copper, and some form of insulation. To manually disassemble the unit and separate the dissimilar materials is time-consuming and therefore costly. To discard these units in landfills is a waste of our natural resources.

2. Changes Made Possible Because of Cryogenics

Through lowering the temperatures of the unit to a −90°C (−130°F), the steel becomes brittle. When subjected to the impacts of hammers in a mill, the cooled steel shatters (Figure 8). Magnets easily lift the steel particles from the mixture. The insulation is separated by air flotation leaving clean, pure copper with 100% recovery.[12]

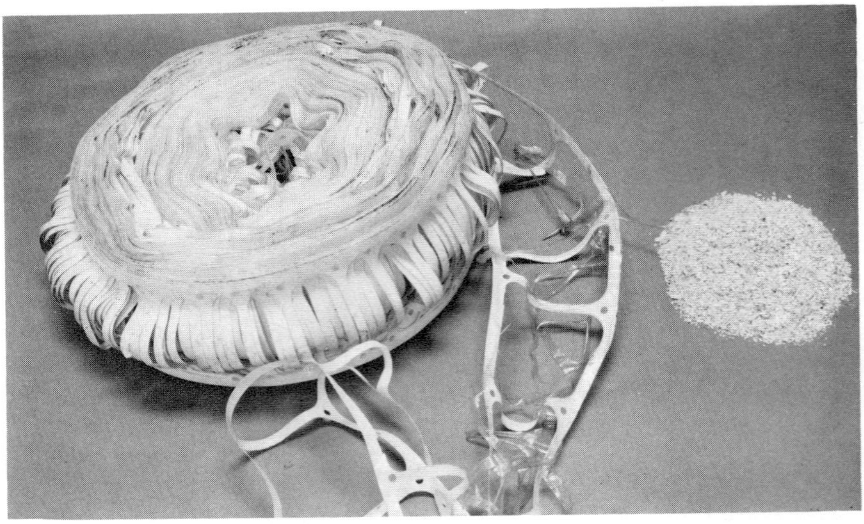

FIGURE 7. Thin film plastic waste from meat packaging industries can be easily and quickly cryogenically processed in preparation for molding into new products.

FIGURE 8. Automobile generators, when cooled to a minus −90°C (−130°F) and subjected to impacts, are fractured into pieces, thereby making simple separation of the materials possible.

3. Advantages and Disadvantages

The greatest advantage is complete separation with 100% recovery for each material through automation.

A disadvantage is that, at this writing, no mill has been specifically designed for cryogenic processing. Therefore current operations are less effective, economical, and efficient.

4. Acceptance

Some secondary metal companies now include cryogenics as a part of their processing plants.

VII. REMOVAL OF PAINT BUILD-UP FROM PAINT-CARRYING HOOKS

A. Case History

1. Problem

Hooks designed to repeatedly carry manufactured items through paint spray booths accumulate layer after layer of paint, often building up to thicknesses exceeding 7.5 cm (3 in.). Eventually it becomes necessary to remove these many layers of paint if the hooks are to be continued in operation. Current practice for removing the paint build-up is by soaking the coated carrying hooks in hot caustic solutions over periods of several days. This is time-consuming, therefore costly, and often does not meet the environmental laws.

2. Changes Made Possible Because of Cryogenics

Through cooling, the accumulated layers of paint become brittle. Impacts from the blows of a hammer or other such devices shatter the paint, which separates, leaving a cleaned hook (Figure 9). This process is fast and efficient, taking only seconds instead of days.[13] Several companies are using this process to clean their carrying hooks. For further information, see Appendix VII.

VIII. SUPERCONDUCTOR MAGNETIC ENERGY STORAGE UNITS

A. Case History

1. Problem

At ambient temperatures, electrical conductors have the property of resistance to the flow of electric currents. Therefore, some of the electrical energy in the system is converted to heat in overcoming this resistance. Also, electric current at ambient temperatures is not conveniently and economically stored in large quantities. It becomes necessary, therefore, to build and maintain power plants able to supply current requirements at peak times, necessitating idling or shutdown of the generators for more than 50% of the time.

2. Changes Made Possible Because of Cryogenics

At certain cryogenic temperatures, specific metals exhibit little or no electrical resistance allowing current to flow uninhibited.

3. Advantages and Disadvantages

It will reduce the cost of building and maintaining the number of power plants necessary to meet peak loads.

A football stadium-sized magnetic coil designed by engineers at the University of Wisconsin—Madison may soon help power plants cut costs, increase efficiency, and reduce the likelihood of blackouts. The huge magnet would be cooled by refrigeration units to hundreds of degrees below zero and buried deep under ground to store excess electricity generated during low consumption periods.[14]

A reported advantage of this system of superconductive magnetic energy storage (known as inductor convertor or IC units) is that the costs per unit of energy go down as the unit gets larger. Also, the environmental impact of the IC unit would be minimal because it produces virtually no noise, dust, or smoke.

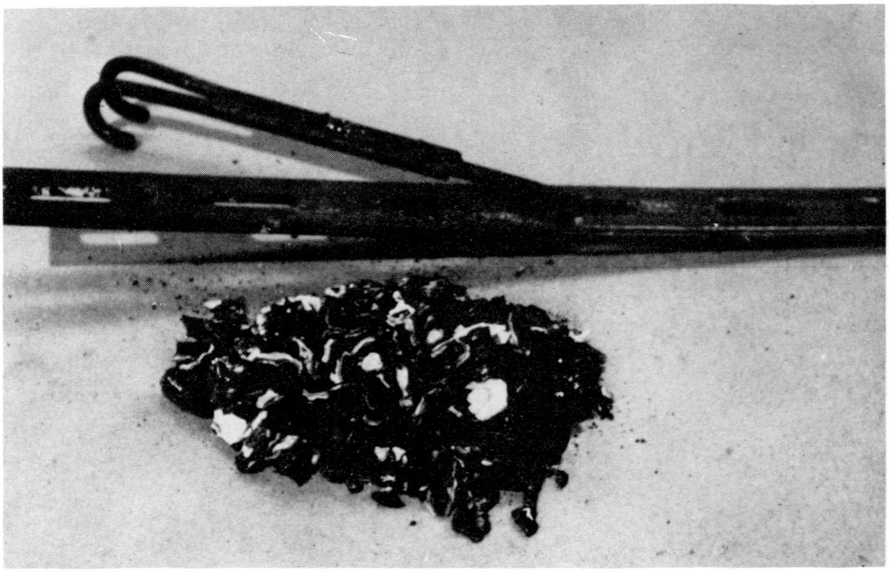

FIGURE 9. Paint build-up on paint-carrying hooks can be removed in seconds when using cryogenic technology.

The magnet would be situated underground to make use of bedrock as an economical support structure because the coils of the magnet have a tendency to stretch when powered by the conductor

Powering the refrigerator to keep the coils cold would be the major operating cost of the IC unit. It is believed, however, that the extra energy provided by the system would more than compensate for the energy consumed in operation.

Realistically, commercial application of IC units is at best several years away. An optimum-sized unit would cost an estimated $200 million to build and would store about 10,000-MWh of electricity, enough to meet the needs of the entire state of Wisconsin. That works out to between 1 and 1.8¢/kWh of electricity for taking generated power, storing it, and returning it to the system.

IX. FAST FOOD FREEZING

A. Case History

1. Problem

Food freezing by mechanical refrigeration is slow, results in dehydration, and consequently a loss of weight and volume. Mechanical devices are expensive to buy, are bulky, and require around-the-clock maintenance and supervision.

2. Changes Made Possible Because of Cryogenics

By using a cryogen (liquid nitrogen), food is frozen in seconds thus preserving much of its original taste, color, and texture.[15] It is reported that weight losses can be reduced considerably when food is frozen cryogenically rather than by other means. Also, cryofreezers are simpler in design and operation than mechanical freezing devices. This reduces cleaning and maintenance hours as well as offering lower capital investment. The system is simple. A stainless steel mesh conveyor belt carries the product through three zones of the freezing tunnel: a precooling zone, a spray zone, and a temperatureequilization zone. Use of three separate cooling and freezing zones provides the most efficient use of cryogenic gas and helps preserve the original appearance, flavor, and texture of the product.

The precooling zone begins at the entrance end of the machine and extends under fans which circulate nitrogen vapor to precool the product for maximum refrigeration efficiency. In the spray zone, droplets of cryogenic liquid vaporize on contact with the product and cause very rapid deep freezing which seals in product quality.

Just before leaving the freezer the product passes through the equalization zone. Here, fan-circulated nitrogen vapor speeds temperature equalization between the product and the tunnel environment for maximum thermal efficiency and product-temperature stabilization.

To accommodate a wide range of production rates as well as varying product characteristics (water content, thickness, belt-loading density, etc.), the freezers are equipped with flexible automatic temperature control systems, a variable conveyor speed, and fan controls. Temperature controls may be mounted for operation from either side of the freezer, and the main control panel may be mounted remotely.

3. Acceptance

Since Mr. Birdseye first experimented with food freezing in the mid-1930s, the process has never been forgotten. During the 1970s the food freezing business has grown by leaps and bounds. Today such large franchises as McDonald's and Kentucky Fried Chicken use cryo-frozen foods exclusively. The economics are elaborated upon in Appendix VIII.[15]

X. REMOVAL OF CORN KERNELS

A. Case History

1. Problem

The kernels of soft sweet corn, when removed from the cob by conventional techniques, are often sliced, losing a quantity of their milk and the germ.

2. Changes Made Possible Because of Cryogenics

The kernels of soft sweet corn can be quickly hardened through an exposure to a cryogen. While in the hardened state the cob with the hardened kernels is passed through a conventional sheller machine. The kernels intact are separated from the cob (Figure 10).

3. Advantages and Disadvantages

These are the advantages of processing young corn at cryogenic temperatures:

1. It appears that frozen soft corn can be shelled quickly at rates similar to that of dried corn.
2. None of the milk from the kernel is lost, as is the case when conventional techniques are used.
3. The "germ" is recovered and retained with the kernel — and important fringe benefit in view of the increasing interest in "natural foods".
4 The reduced temperature used for removing the kernel is also a step toward final processing the kernel as a frozen packaged food. One freezing achieves two purposes.
5. The undesirable hull is easily separated from the kernel as it is removed from the cob.

FIGURE 10. The use of cryogenics for the removal of corn kernels from the cob while in milky state.

4. Acceptance

Canneries and cannery research centers are evaluating the cryogenic technique.

Related U.S. patents are (1) 3,070,447—Webster—Method of Treating Corn, (2) 2,085,051—Taylor—Treating Green Corn.

It is reported that by weight, approximately 75% of the corn picked is waste. After processing, this waste (husks and cobs) must be returned to the field or to some other area for disposal. This is time-consuming and costly. The development of a process whereby the cob could be husked, frozen, and shelled while still in the field could reduce trucking and labor costs dramatically. The husks are natural soil conditioners and fertilizers. If fragmented, the cob is equally as good.[16]

XI. TOMATO PEELING

A. Case History

1. Problem

There is a 12.7% skin and trim loss with commercially peeled tomatoes using the boiling water technique. This figure could easily escalate when the fruit is overripe.

2. Changes Made Possible Because of Cryogenics

The skin and trim loss is reduced to 6.8% when the tomato is crust-frozen.

3. Advantages and Disadvantages

Cryogenically peeling tomatoes saves time, reduces waste, and improves their visual color.

4. Acceptance

A greater detailed discussion on liquid nitrogen tomato peeling is found in Appendix IX.[17,28]

XII. CRYOSURGERY

A. Case History

1. Problem

Surgery at ambient temperatures involves any number of problems, such as the interference of blood and other fluids. The possibility of infection developing from surgery is also of constant concern. Performing surgery at cryogenic temperatures can play an important part in reducing these problems.

2. Changes Made Possible Because of Cryogenics

Pain can be reduced by lowering the temperature of a particular part of the patient. Hemorrhaging can be slowed or eliminated, and preservation of the tissue can be improved.

3. Advantages and Disadvantages

History records that the use of cold to heal wounds is not new. In fact, in 3000 B.C. this practice was not uncommon. Since the early 1960s, cryogenic surgery has developed rapidly, and today is an accepted surgical tool. In fact, with modern equipment and technology, cryo-treatment can be performed with pinpoint accuracy.

4. Acceptance

Cryosurgery is being practiced today in areas such as opthalmology, urology, neurology, and gynecology.[18] The treatment of Parkinson's disease is perhaps one of the more commonly used applications. Fine, specially designed needles can be accurately placed within the critical section of the brain. Cataracts are also removed using the freezing process. The lens is frozen, and once hardened, can be separated from the eye. Other common uses of cryosurgery are the removal of tonsils and warts and treatment of tumors and defective prostate glands. As the science for cooling and rewarming the body cells becomes better known, the list of applications of cryosurgery will undoubtedly expand enormously.

XIII. PRESERVING OF BIOLOGICAL SPECIMENS

A. Case History

1. Problem

Over the years of selected crossbreeding, desirable characteristics can be developed within the mammal only to be lost when the life cycle ends.

2. Changes Made Possible Because of Cryogenics

Without low-temperature storage, there is no known technology to preserve semen.

3. Advantages and Disadvantages

With the advent of cryogenics, it is now possible to freeze the semen from the choice male mammal for impregnation of a female years later.[19]

Semen collected from males containing desirable features or characteristics can be cryogenically stored and used years later to rejuvenate these features.

4. Acceptance

In 1978, nearly 50% of the dairy cows in the U.S. were bred by cryogenically preserved semen. It is predicted that by 1985 this figure will rise to nearly 100%. Much work is currently in progress for the cryogenic preservation of vital organs and limbs of humans for the purpose of using them at some later date. In July of 1979 a person's hand was completely severed in an industrial accident. The hand was sealed in a plastic bag and then cooled. Hours later, at the University of Wisconsin hospital, it was rejoined to the original arm, and within 4 days the person could once again move the fingers.

XIV. SLAUGHTER HOUSE APPLICATIONS OF CRYOGENICS

A. Case History

1. Problem

To commercially remove the hair from slaughtered hogs not only involves several steps but also the soaking of the entire hog in a hot solution. These procedures ruin the hide for high-quality uses such as leather products. In the past, and in most instances even today the main use of pigskin was as a gelatin product (Figure 11).

2. Changes Made Possible Because of Cryogenics

By cooling the outer surface of the skin for a slaughtered hog to −68°C (−90°F), the hair can be fractured at its roots using a circular motion of a nylon bristle brush.[20]

3. Advantages and Disadvantages

By using the technique of cooling rather than cooking, the properties of the hide are not changed and therefore are available for use as high-quality leather products.

4. Acceptance

This process has been limited to laboratory experiments.

XV. CRYONICS

A. Case History

1. Problem

Today the world is filled with people who have incurable diseases and expect to deteriorate and die before a cure can be found. There are people who have the desire to risk a few dollars of their estate with their bodies placed in cryonic suspension with high hopes of being physically revitalized at a more appropriate time.

2. Changes Made Possible Because of Cryogenics

With the advent of cryogenics, "cryonics" is becoming a household word. Cryonics is defined as the act and embryonic science of freezing dead human beings until the day when medical science is capable of coming to the rescue.

3. Advantages and Disadvantages

The advantage at this time, it seems, is "hope" for those who believe. The disadvantage is the annual cost of approximately $2000 after the initial fee of $10,000 has been paid. Common practice has been that relatives, finding what they believe to be a more appropriate use for the $2000 annual fee, have ignored the will. The body is then moved to a grave. To overcome this loss of confidence and "will-breaking" procedure, one who is interested in cryonics can now take out an insurance policy which would guarantee one's wishes.[21]

4. Acceptance

There are no compiled records available as to the exact numbers of human cadavers

FIGURE 11. The use of cryogenics in the slaughter house. By cooling the skin of freshly slaughtered hogs to −68°C (−90°F), hair shatters away from the skin when rubbed with a nylon bristle brush.

in cryonic suspension. It is reported that perhaps two dozen or more bodies were placed in cryonic capsules in the 1970s. At one time cryonic facilities were reported available in the cities of San Francisco, Los Angeles, New York, and Phoenix.

XVI. BOMB DISPOSAL

A. Case History

1. Problem

Highly trained, bomb recovery teams routinely risk their lives when assigned to defuse and remove bombs. For example, in 1978 the habitants of a house in Krasnormeysh, Ukrainian Soviet Union found a 250-kg bomb buried deeply below the floor of their home. This bomb became lodged there in 1943 after piercing the roof and floor but did not explode. By reducing the temperature of the detonator to −90°C (−130°F), the spring in the component lost its ability to spring, and the bomb was safely removed.[22]

2. Changes Made Possible Because of Cryogenics

Detonators are activated by a spring mechanism. By lowering the temperature of a steel spring to −90°C (−130°F), its embrittlement temperature, the detonator becomes inoperative and is in stable condition for removal and proper disposal.

3. Advantages and Disadvantages

Using a portable Dewar with a 5-psi head pressure, liquid nitrogen can be easily sprayed onto the detonator.

Military wars and civilian riots are both causes for abandoned bombs which, when found, require the attention of bomb-disposal squads. The cryogenic process could be helpful in preventing loss of life and injury, as well as effecting a possible reduction in the ever-rising insurance premium on the lives of the disposal teams.

XVII. RECOVERING WATER-SOAKED BOOKS

A. Case History

1. Problem

From time to time valuable books are subject to floods which are almost certain to destroy the page as well as the print. Also, the wet pages will readily mold, mildew, distort, and adhere to other pages.

2. Changes Made Possible Because of Cryogenics

Water-logged books can now be placed in a vacuum chamber at cryogenic temperatures. Under these conditions the water turns to ice crystals, and because of the high vacuum, the ice crystals are vaporized, leaving the previously wet pages dry and brittle. When these books are placed in a high-humidity environment under controlled conditions, the proper amount of moisture can be returned to the pages.

3. Advantages and Disadvantages

This rather recent application of cryogenics makes possible the safe salvage of books with "collector item" values.

4. Acceptance

The above salvage operation was successfully performed at Stanford University when, in November 1978, approximately 30,000 books became soaked when a water main broke and flooded the library.[23] The books were rejuvenated and replaced on the shelves in April 1979.

XVIII. ELIMINATION OF "BOOKWORMS"

A. Case History

1. Problem

It is not uncommon to find "boring beetles" nibbling away at the pages and bindings of older books on the shelves of our libraries. It takes very little imagination to visualize the results should this practice be permitted to continue.

2. Changes Made Possible Because of Cryogenics

Entomologists have found that putting the bug-infested books through a freezing cycle will kill the bugs. It is reported that temperatures as low as $-30°C$ ($-22°F$) for a predetermined length of time is adequate.

The system has been tried in the library at Yale University and has generated interest at other libraries, including the Library of Congress.[24] The technique is quick, efficient, and safe. A newspaper reports that Yale University has completed its task with a degree of success of freezing 37,000 volumes of bug-ridden books.

XIX. CRYOGENS USED TO PROPEL VEHICLES

A. Case History

1. Problem

Whereas gasoline and diesel fuels have served the average person well in propelling his automobile or truck for the past three quarters of a century, the same fuels would present problems if used in the space program.

2. Changes Made Possible Because of Cryogenics

Dating back as early as 1926, Dr. Robert Goddard used a cryogen (liquid oxygen) to propel his first rocket. Thirty years later, in the 1950s, cryogens such as liquid nitrogen and hydrogen served several functions, including propulsion in the well-known space program. In the 1970s, liquid helium was added to the space cryogen list. For example, for the Apollo XI moon flight alone, more than 8900 metric tons (10,000 tons) of liquid oxygen and nitrogen were used; 44,640 metric tons (50,000 tons) of liquid oxygen and nitrogen were used for the Apollo flights VIII through XI.[25]

In view of present-day gasoline shortages, researchers are taking a look at cryogen-propelled ground vehicles.

XX. CONCLUSION: CRYOGENICS IN THE YEAR 2000—A PREDICTION

The art and science for cryogenics has been widely accepted by processors internationally. Its popularity and growth can easily be compared to that of the steel melting furnace.

Two hundred years ago the furnace was in its infancy, very close to where cryogenics is today. During this period the furnace has had its ups and downs, and has gone through several changes. Like the furnace, cryogenics will be attempted for many applications — some will fail while others will be successful. A recent comment in *Reclamation, Conservation and Beautification*[26] refers to cryogenic technology as one dramatic advance in the processing industry: "The scrap processing industry continues to work at ways of improving its products and its ability to meet the needs of its customers. One dramatic advance has been the adaptation of cryogenics processing; metals are dipped in a bath of liquid nitrogen which freezes them to extremely low temperatures — around −184.44°C (−300°F). At these temperatures metals become brittle and can be shattered into small sections for easier separation. Cryogenics is ideal for the processing of complex items such as electrical motors which contain a variety of metals — all valuable — but for which hand separation was the only way of separating the components. Experiments continue to find ways of applying this technique and others to the scrap processing industry."

Not all materials become embrittled at reduced temperatures; therefore, such materials, when used alone in a product, may not be cryogenically recyclable. Fortunately, however, most products are made from a combination of materials of which one or more are often affected by cold temperatures, making selective cryogenic processing possible.

An example of this is the processing of copper wire insulated with a plastic coating. The plastic becomes brittle at reduced temperature, and the copper does not.

It would appear that within one decade the technology necessary for using cryogens has advanced to the stage where it is now appropriate to develop the markets for the recycled materials. As the demands for the materials increase, so will the interests and investments of industries. The rapid increase in the use of rubber in roads, streets, and runways is an example. Only 2 years ago, the U.S. Department of Transportation offered financial reimbursement to government agencies interested in laying stretches of rubberized asphalt. This was known as Project 37. At last count, more than 30 states were participating.

It is because of motivating projects such as Project 37 that today a U.S. cryogenic gas-producing company has licensed four tire-processing companies (one in Holland, one in England, and one each in Indiana and Ohio) to use their patent in cryogenically generating rubber particles from used tires. Cooling facilities for purposes other than for the processing of used tires are becoming more and more common, not only in the food industry and medical centers, but also in industries such as the foundries, machine shops, and secondary metal plants.

Figure 12 is a flow chart showing how several materials can be processed, some cryogenically, prior to their reuse.

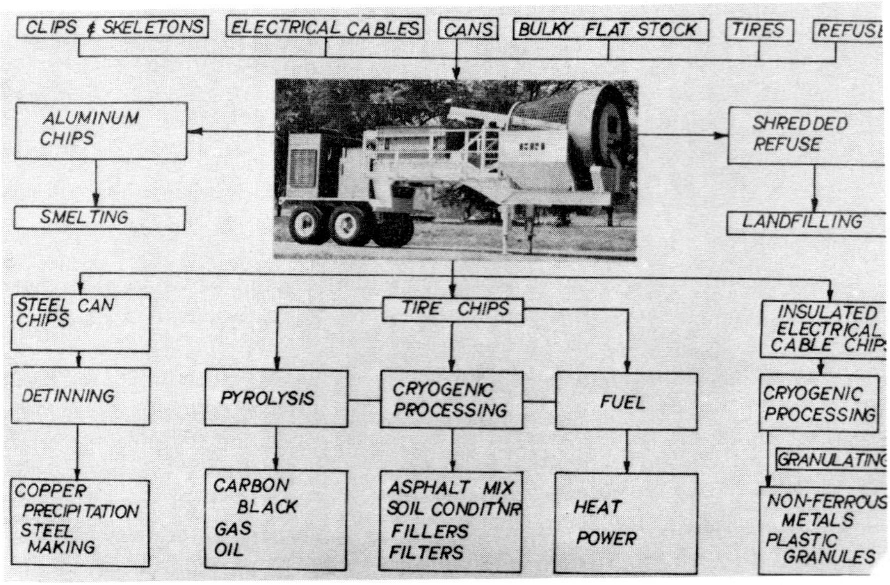

FIGURE 12. Schematic of the usefulness of cryogenics in recycling of materials.

Cryogenic technology is well known and public interest and curiosity is high. The principle is environmentally sound. The cryogens come from the atmosphere, are used to cool or embrittle, and are then returned intact.

REFERENCES

1. TIREX Corp., Southlake, Tex., personal correspondence, 1977.
2. **Costly rubber street may quiet rail crossing,** *Capital Times, (Madison, Wis.,)* p. 57, September 14, 1977.
3. **New technology solves bridge problem,** *Public Technology,* Washington, D.C., February/March, 1976.
4. **Nickerson, W. J. and Faber, M. D.,** Soil Conditioner, Firestone Rubber Co., U.S. Patent 3,766,685, 1973.
5. **Locketz, H. B.,** Fuel Composition, Cryogenic Recycling International, U.S. Patent 3,864,094, 1975.
6. Humane Equipment Corp., Baraboo, Wis., personal correspondence, 1979.
7. **Braton, N. R.,** The Use of Cryogenically Processed Rubber in Oil Spill Clean-Up Operation, College of Engineering, University of Wisconsin, Madison, 1979.
8. **Tharon, D. W.,** Mercury Removal From Aqueous Process Streams, E. I. duPont de Nemours, U.S. Energy Commission report, Savannah River Laboratory, Aiken, S.C., January, 1974.
9. Municipal Environmental Research Laboratory, Office of Research and Development, Tires: Decreasing Solid Wastes and Manufacturing Throughout; Markets, Profits and Resource Recovery, EPA-600/5-78-009, U.S. Environmental Protection Agency, Cincinnati, Ohio, July 1978.
10. **Haberman, C. E.,** Pyrolysis of Scrap Tires Using the TOSCO II process, Oil Shale Corp., Golden, Colo., September 1, 1976.
11. **Tam, K. W.,** Recycling of Thermoplastic Films, M.S. thesis, College of Engineering, University of Wisconsin—Madison, 1977.
12. **Roman, G. C.,** The Design of a Cryogenic Recycling System, M.S. thesis, College of Engineering, University of Wisconsin—Madison, 1973.

13. **Braton, N. R.,** Removal of Built Up Layers of Organic Coatings, College of Engineering, University of Wisconsin—Madison, U.S. Patent 3,934,379, 1976.
14. **Boom, R. W., Peterson, H. A., and Young, W. C.,** A look at superconductive storage, *Electrical World,* March 1975.
15. Union Carbide Corporation, Linde Division, Food Technology and Cryogenic Chiller Systems, Bull. 0756-5MF-4177, and F3962, 3M-86-9414, Linde Division, Union Carbide Corp., N.Y., October 1977, June 1976.
16. **Braton, N. R.,** The Use of Cryogenics For the Removal of Corn Kernels While in a Milky State From the Cob, College of Engineering, University of Wisconsin—Madison, 1978.
17. U.S. Department of Agriculture, Liquid Nitrogen Tomato Peeling, U.S. Department of Agriculture Food Crops Utilization Research Laboratory, Weslaco, Tex., 1978.
18. Union Carbide Corporation, Linde Division, Cryosurgery, Brochure 3300-P89-1097, Linde Division, Union Carbide, New York, 1978.
19. Union Carbide Corporation, Linde Division, Preservation of Biological Specimens, brochure, Linde Division, Union Carbide, New York, 1978.
20. **Braton, N. R.,** Cryogenics Used in the Slaughter House, College of Engineering, University of Wisconsin—Madison, 1978.
21. **Williams, G.** Deep freeze for the dead in Emeryville, *San Francisco Chronicle,* August 23, 1977.
22. **Benyukh, O. P., Ed.,** Cryogenics used in bomb disposal, *Soviet Life Magazine,* p. 29, March 1979.
23. Cryogenics used to restore/recover water logged books, *State Journal, (Madison Wis.,)* March 25, 1979.
24. The use of cryogenics to eliminate book-eating bugs from rare editions, *Capital Times, (Madison, Wis.,)* July 1979.
25. Union Carbide Corporation, Linde Division, Cryogens Used to Propel Vehicles, Brochure F-3300, Linde Division, Union Carbide, New York, 89.
26. Institute of Scrap Iron and Steel, Reclamation, Conservation and Beautification, Institute of Scrap Iron and Steel, no date.
27. **Kisielewski, R.** Flow chart for processing materials, *Cryogenic Recycling International,* LaCross, Wis., 1977.
28. Liquid Carbonic Corporation, Liquid Nitrogen Tomato Peeling, Tech. Sales Bull. No. IA-1072, Liquid Carbonic Corp., Chicago.

APPENDIX I

CRYOGENIC PROCESSING OF THE TIRE*

I. INTRODUCTION

Each year, between 160 and 200 million tires are scrapped in the U.S. Most of these tires are literally thrown away, wasting a tremendous source of valuable recyclable material. Conservatively, these tires contain 1.5 billion kg (3.3 billion lb) of rubber that could be available for recycling, if they could be economically processed. Of the tires scrapped each year, 40 million tires are retreaded, 14 million are reclaimed, and 6 million are used in miscellaneous ways. This still leaves 1 billion kg (2.2 billion lb) of rubber lost each year and a 100 million tire-disposal problem.

Some form of rubber recycling has been practiced for over 120 years. In the tire industry, the basic methods for recycling tires are chemical devulcanization and mechanical grinding. However, the advent of the steel-belted tires, combined with increasing power, labor, capital, and maintenance costs, have made it increasingly difficult to process tires with conventional methods. Recent development work utilizing cryogenic technology suggests that a liquid nitrogen-based system offers the best solution for processing the various types of tires scrapped in the U.S.

During the last 1½ years, a demonstration facility has been assembled and operated for the purpose of developing and testing a cryogenic tire-fragmentation process. The testing has been completed, and quantitative analysis has shown that a crumb rubber of high purity can be economically and efficiently produced using cryogenic technology.

The purpose of this paper is to report on the development of this system as an alternative method for processing scrap tires into valuable recyclable materials.

II. TECHNOLOGY

Cryogenic technology is not new to the rubber industry. The initial work done in this area proved that it was technically feasible to freeze, pulverize, and recover rubber in crumb form, but most of the early work did not efficiently utilize the liquid nitrogen coolant. This resulted in relatively high liquid nitrogen consumption and a less-than-attractive economic picture. However, recent technological developments in cryogenic freezing have significantly reduced the liquid nitrogen consumption, a major factor in making cryogenic recycling of scrap tires economically viable.

In order to obtain data with a high degree of reliability, the facility had to be large enough to supply accurate processing information. This would provide a high degree of confidence in the data and allow a more accurate scale up to a production-size facility. The facility has the capacity of processing 454 kg (1000 lb)/hr or approximately 50 tires per hour. This system has enabled the accumulation of accurate data and at the same time provided a source of clean crumb rubber for end-use product development.

A conceptual flow schematic for the process is shown in Figure 1. It is based on U.S. Patent 4,025,990.

* From Kazarnowicz, M. C., Osmundson, E. C., Boyle, J. F., and Savage, R. W., Cryogenic Scrap Tire Processing, paper presented by M. C. Kazarnowicz at the 112th meeting of the American Chemical Society's Rubber Division, Cleveland, Ohio, October, 7, 1977.

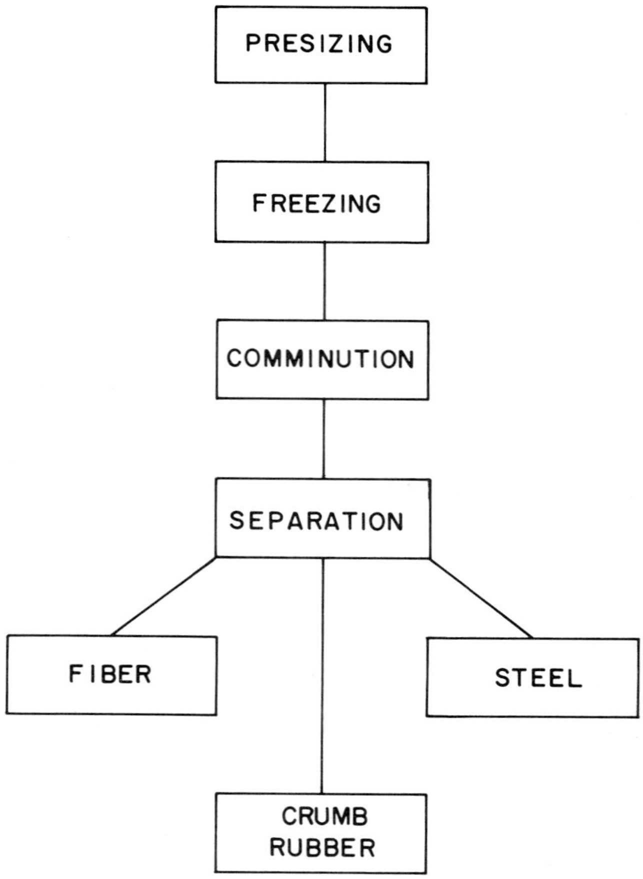

FIGURE 1. Schematic of recovery of rubber in crumb form.

The first step is presizing. In this step, the tires are reduced to pieces approximately 15 cm^2 (6 in^2.) This is necessary to reduce the overall size of the input and to provide an increased bulk density of the tire sections being fed into the freezer. This increased bulk density helps to optimize the liquid nitrogen consumption, one of the factors in reducing the overall process costs.

The heart of the system is the freezing tunnel, specifically designed for processing tires. The tire sections are cooled to the point of embrittlement; this is generally between −60°C (−76°F) and −100°C (−148°F). At these temperatures the rubber passes through glass transition and becomes brittle. In this condition, the rubber can be easily liberated from the metal bead and tire cord, as small particles.

Liquid nitrogen is used as the cooling agent. As a liquid at cryogenic temperatures, it is easy to control and has a high cooling capacity relative to its cost. The liquid nitrogen is injected into the freezer so as to maximize the heat transfer between the nitrogen and the scrap tires. This is accomplished by utilizing the boiling heat transfer in a liquid nitrogen spray zone and the convective heat transfer in an extended fan-circulation zone. In the spray zone, the liquid nitrogen is sprayed directly onto the surface of the scrap tires. The liquid, upon contacting the tire sections, is instantly vaporized, absorbing its heat of vaporization from the material being processed. The cold nitrogen gas is then recirculated by a series of fans which increase the nitrogen gas velocity. This increased gas velocity enhances the convective heat transfer effect, thereby improving the overall freezer efficiency. Having passed through the freezer, the material is then comminuted in the size-reduction mill, and the rubber is liberated from the steel and fiber in a single operation.

After exiting from the size-reduction mill, the liberated material can be magnetically separated and screened into the various product streams.

III. PROCESS CHARACTERISTICS

There are many advantages to using a cryogenic system for recycling scrap tires. One of the principal advantages is the flexibility inherent in its design. The product streams generated can be easily changed to accommodate specific product requirements simply by changing any one of a number of interacting variables. The operating temperature of the freezer and the retention time in the size reduction mill are two of the most important operating parameters which can alter the final product characteristics significantly.

For example, the following graph demonstrates the effect operating temperature has on the product streams. Figure 2 shows the particle size distribution as a function of operating temperature. As the operating temperature decreases from −35°C (−67°F) to −100°C (−148°F), the percentage of the −4.0 mm product increases from about 23 to 50%. With this type of information, the system can be tailored to meet various particle size requirements. Data on these, and other operating parameters, have enabled the system flexibility to be quantitatively defined. This data base can now be used to pinpoint the operating conditions which yield a specific product stream, thereby optimizing the product outputs and economics of the entire system. For example, rubber yield can be traded off against rubber purity, or smaller particle size product can be obtained at the expense of an increased operating cost.

Additional advantages of a cryogenic system include the following:

1. The capability to recover a high percentage of the available rubber. Generally, 90 to 95% of the rubber is recovered with a cryogenic system.
2. The production of a crumb rubber that is virtually free of fiber and metal contaminants.
3. The ability to process both fabric and steel-belted tires without bead removal. This eliminates the costly and wasteful bead-removal step necessary in some reclaiming operations. It has been estimated that as high as 8% of the available rubber in a tire can be lost in this operation.
4. The ability to produce various mesh size particles by varying the operating parameters.
5. A relatively simple system design with considerably lower horsepower requirements, as compared to noncryogenic systems. This will allow for lower maintenance and power costs. Connected horsepower requirements can be up to 50% lower than noncryogenic systems.
6. The possible sale of high-purity steel which can be recovered.
7. The possible sale or use of fiber as a fuel in waste heat boilers.
8. Low capital investment.

IV. PRODUCTS

The crumb rubber is virtually free of fiber and steel contaminants. It is composed of distinct rubber particles. No fiber or steel filaments are entrapped in the rubber. This is a result of the complete liberation obtained on impact of the embrittled material in the size-reduction mill.

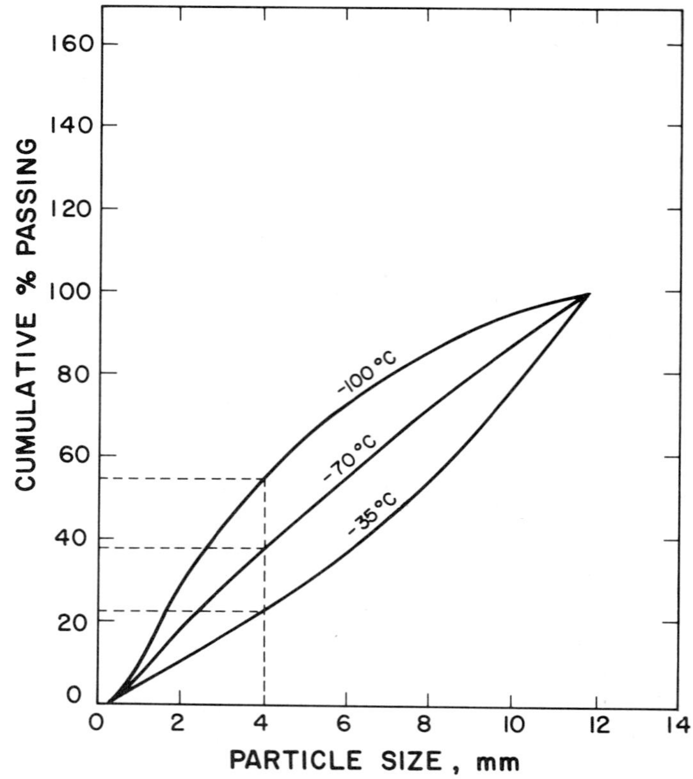

FIGURE 2. Particle size distribution as a function of operating temperature.

The rubber recovered by the basic system would generally range in particle size from 0.8 mm (20 mesh) to 8 mm (2 mesh), although most particle sizes can be produced by altering the operating parameters of the system or adding subsequent grinding steps.

The steel obtained is contaminated with fiber and a very low percentage of rubber. Preliminary tests have shown that a relatively high-purity steel can be obtained by burning the contaminants from the steel. This suggests that the metal can be processed and recovered by low-grade smelting operations.

The recovered fiber contains rubber of various particle sizes which were trapped by the "nesting" effect of the fiber during separation. One potential use that has been suggested is burning the fiber to recover its heat value.

Again, although a small amount of rubber is found in the by-product streams, the overall rubber recovery can be maintained at an economically high level.

V. ECONOMICS

The true test of a particular process lies in its operating economics. During the developmental phase of this program, equal emphasis was placed on process economics as well as process capability. As is the case with all cryogenic systems, the liquid nitrogen consumed represents a sizable portion of the operating costs. For this reason a significant effort was directed toward freezer design in an attempt to minimize this consumption as much as possible.

Process economics have been developed for a typical cryogenic scrap tire recycling facility using the data obtained from the demonstration plant. Table 1 shows the esti-

Table 1
ESTIMATED CAPITAL INVESTMENT AND OPERATING COSTS

Cost of operation	$/year	$/kg
Depreciation	35,000	0.31
Interest	17,500	0.16
Power	26,250	0.23
Labor	77,000	0.68
Maintenance material	35,000	0.31
Liquid nitrogen	628,250	3.89
Total	819,000	5.58

Note: Basic operating equipment: $350,000; Capacity: 11,250,000 kg of scrap tires per year

mated economics for a typical facility processing at least 11,250,000 automobile tires a year, or about 11,250,000 kg (24.8 million lb). These economics are based solely on the cost of the basic operating equipment necessary in the production facility and do not include: the manpower to load or package the product, conveyors, and dust collection equipment or installation. These would vary for each facility and were therefore excluded.

The $5.58/kg ($2.53/lb) figure is the operating cost to obtain a 13 mm (0.5 in.) and smaller crumb rubber product.

VI. END USES

There are a number of areas where crumb rubber can be used. A range of molded and extruded rubber products is currently being produced. These products include: flower pots, irrigation hose, garbage cans, fence posts, railroad ties, and pallets. In addition, there are a number of research programs currently under way which are investigating potential end uses for the crumb rubber. These programs are investigating end uses in the plastics, asphalt, and tire retreading industries.

VII. SUMMARY

Cryogenic processing of scrap tires now offers the rubber industry a viable method for recycling the millions of tires scrapped each year. The many advantages of a cryogenic recycling system include: production of a rubber product virtually free of fiber and steel contaminants, the flexibility to produce almost any particle size, the ability to process both fabric and steel tires, and the recovery of a high percentage of the available rubber with a relatively simple system design. This system demonstrates an economical means for recovering the vast quantities of rubber and at the same time reducing the overall solid waste disposal problem in the U.S.

APPENDIX II

REPORT OF TEST*

Client: Safe-T Blade, c/o Prescott Valley, Inc., 6040 North 7 Street, Suite 101, Phoenix, Ariz. 85014

Subject: Evaluation of Nylon Lawn Mower Blades

Authorization: Client's Order for Test dated April 20, 1976.

Specification: American National Standard No. B71.1a-1974, "Safety Specifications for Power Lawn Mowers, Lawn and Garden Tractors, and Lawn Tractors".

Sample Identification: Six (6) samples of the client's nylon lawn mower blades sampled and submitted for testing. The samples were 21 9/16-in. long × 2 3/16 in. wide × ¼ in. thick, with the first 4½ in. of each leading edge chamfered approximately 25° to produce a cutting edge.

The tests were performed with the sample blade mounted on a 21 in. Toro "Whirlwind" lawn mower procured by the laboratory at the client's direction.

Conclusions: The Safe-T Blade lawn mower blade samples were found to comply with all the blade-related requirements of the referenced ANSI standard when tested as described herein.

Test Procedures and Results: The lawn mower with the "Safe-T" blade installed was subjected to the tests prescribed in the listed applicable paragraphs. The tests, their requirements and results are tabulated below:

Sound Level Test** — max, if 82DBA at points 10 in. each side of C/L, 8 in. behind handle, 66 in. above ground.
Results: With original blade — 96DBA; With Safe-T blade — 93.5DBA

Sudden Impact Test — any failure of blade or blade-retaining device shall be considered a failure.
Results: No failure of blade or blade-retaining device was noted after the test.

Unbalance Test — mower shall run 1 hr with the blade unbalanced by $L^3/1855$ inch-ounces without loss of any part or failure of any component.
Results: No parts were lost nor any components failed during the 1-hr test.

Blade Security Test — blade shall resist 50 ft·lb torque trying to rotate it relative to the mower shaft.
Results: Blade withstood 68 Nm (50 ft·lb) torque without moving relative to the mower shaft.

Structural Integrity — after introducing 100 hardened steel bolts (½ in. diameter) into the rotating blade, there shall be (1) no penetration of more than one half of the ball through any surface, or (2) the breaking off or bending of any part of the housing guard or grass catcher.
Results: No such damages resulted.

Thrown Object Test — Discharge Opening — not more than 15% (90) of the 600 nails introduced into the mower shall strike the target above the horizontal plane of the blade, and not more than 0.5% (3) shall strike the target above a line 24 in. above the blade plane.
Results: 48 nails were within the target area and 2 reached above the 24-in. line.

* Safe-T Blade Co., Phoenix, Ariz., personal correspondence, 1979.

** This comparison was run under less-than-ideal prescribed conditions of the specification just as a matter of interest.

Probe Test — the standard probe shall not contact the blade in either its highest or lowest position when the probe is rotated 15° either way horizontally and vertically.
Results: No blade contact was made.

APPENDIX III

NEW TECHNOLOGY SOLVES BRIDGE PROBLEM*

A report from the Eugene, Ore. has informed us that the new, modern surface on the concrete deck of the Ferry Street bridge there, has proved successful over the past winter. Cracks in the bridge deck had been a problem for city engineers during several years, and none of the proposals for preserving the deck appeared to promise anything like a permanent solution at a reasonable cost. The problem was important because the Ferry Street bridge is the city's busiest and each day carries more than 42,000 autos, buses, and trucks over the Willamette River.

Early in 1975, the city administration asked Greg Page, Technology Agent of the Urban Technology System, to take a fresh look at the problem in conjunction with Department of Public Works Director Don Allen and his staff. Eugene is a member city of the Urban Technology System and as such, has a resident Technology Agent or Technology Coordinator (as he is called there) on its government staff. Greg Page put together a problem statement using the standard PTI method for these initiation papers, and discussed it with engineers at Battelle Northwest Laboratories, the technical backup site contracted by PTI to assist him and the city with problems of this kind.

It was apparent to all that, whatever the choice of technology for a solution, the bridge deck surface treatment had to be tough enough to withstand tens of thousands of tire studs daily in winter. It also had to be flexible enough to allow for expansion and contraction — up to 4 in. between winter and summer. Finally, local considerations restricted the city to technology applications that would allow the bridge to remain open to traffic during all morning and evening rush hours, while the resurfacing was under way.

The city was offered three possible solutions and approved one after Department of Public Works Director (DPW) Don Allen had evaluated them. The solution chosen called for the use of a commercial rubberized asphalt surface with flexible concrete expansion dams at either end of the bridge. A report at the time the choice of technologies was made noted that "rubberized asphalt does not represent a final solution (there is none within the existing state of the art) but it is the most economical based on wear data from similar installations." Mr. Page was able to secure evaluations of this material through the Urban Technology System network.

The concrete expansion dams are a new technology developed by Battelle, and made available to Eugene for this project. The technology calls for infusion of high-alumina concrete with small fibers of specified size and quantity. The final product has demonstrated its ability to meet the use criterion (5000 psi during a 3-day compressive strength test) without deterioration, under actual use conditions, for more than 8 months. It has provided a smooth transition between the bridge deck and the land surface despite the natural tendency of the bridge to expand and contract under differing temperatures.

Actual construction work on the bridge project was done in August 1975. Mr. Page worked closely with city crews and with Assistant DPW Director Don Gilman, helping them to maintain tight quality control. The entire job required three straight days of round-the-clock work, with operations timed to keep open half of the bridge's traffic lanes during rush hours.

The city of Eugene has had 8 months to test the new bridge decking, and from that

* Public Technology, Inc., N.W. 1140 Connecticut Ave., Washington, D.C., 20036, Feb/March, 1976.

vantage point, Mr. Allen said recently, ''Both the deck overlay and the expansion dams are holding up exceptionally well. The three-way partnership — our public works staff, Eugene's Technology Coordinator, and the researchers at our Urban Technology System backup site, Battelle Northwest Laboratories — produced optimum solution to a many-faceted problem; that of resurfacing the bridge deck with the smallest possible deadload, at minimum cost, and with the shortest disruption to the community.'' Mr. Allen also noted that the city administration avoided $30,000 in unnecessary costs for the job by following Battelle recommendations.

Greg Page said, ''When I look back on it, I feel really pleased that we were able to do something helpful for the drivers and taxpayers''.

APPENDIX IV

SOIL CONDITIONER*

I. BACKGROUND OF INVENTION

Many millions of acres of potentially arid land throughout the world are not employed because the land cannot efficiently utilize the water which is available either as precipitation or by irrigation. Often the soil has such high clay content that the water cannot be absorbed and it runs off. In other areas the land has such a high sand content that the water immediately sinks into the ground to a depth where it is not available to support vegetation. Huge areas of the world, such as parts of North Africa and the Middle East, once supported agriculture and are no longer able to do so because poor farming practices have laid waste to the land.

These millions of acres could be turned into useful farm land, and the wasted areas could be reclaimed if they could be made to utilize available water efficiently, so as to support vegetation and eventually build up an enriched top soil.

One of the truly serious problems of modern society is the disposal of tire wastes. Old tires cannot be efficiently burned, buried, or regenerated. Destructive distillation has been used to some extent to dispose of tire carcasses, but this procedure is costly, and may be a source of atmospheric pollution. As a result, many millions of old, used tires accumulate in unsightly dumps in areas surrounding our cities.

A method has now been discovered to mitigate both problems — the problem of unused land and the problem of accumulated used tire carcasses.

II. THE INVENTION

This invention specifically relates to soil conditioners produced by fermentation of microorganisms using tire mesh as a substrate. The novel products of the invention are soil conditioners comprising oxygen-enriched particulate fermentation products of used tire mesh integrally bonded to the mycelium produced by the fermentation organism. They can be suspended in water and when dispersed in sandy or clayey soil, will markedly increase the capacity of the soil to retain water. Because of the presence of the organic mycelium, the products themselves are useful nutrients to support vegetative growth. The products also have ion exchange capacity and therefore can be used to enrich the soil with metallic nutrients.

While the invention is particularly applicable to, and of great economic and ecological value with, used tires, it is of course analogously useful with other equivalent rubber sources.

Used tires as a raw materials source comprise a heterogeneous mix of natural and synthetic rubbers containing a wide variety of adjuvants. Normally, tires contain both natural and synthetic rubbers, the relative proportions of each depending upon the type of tire. Truck and other heavy-duty tires contain a high proportion of natural rubbers. Synthetic rubbers employed in tire production include for example, butadienestyrene rubbers, butadiene-acrylonitrile rubbers, butyl rubbers prepared from isobutylene, neoprene rubbers prepared from chlorprene, chlorosulfonated polyethylenes, polysulfide rubbers prepared by reacting sodium polysulfide with chlorine

* Nickerson, W. J., Rutgers University, New Brunswick, N.J.

terminated aliphatic hydrocarbons or ethers, and silicone rubbers which are the polymerication products of diallyl siloxanes, principally dimethyl siloxane. The process of this invention is applicable to all of these rubber substrates, and to others which are sometimes employed.

In the preparation of tires, the rubber will be compounded with a wide variety of other materials. They are vulcanized so of course will contain sulfur and vulcanization aids, particularly accelerators. They will also usually contain antioxidants; antiozanants; various resins, especially of the coumarone-indene type; zinc oxide, carbon black, and hydrocarbon oils of the naphthenic and aromatic class. Typically, truck tires will contain at least 50 to 90% synthetic rubber. The average amount of hydrocarbon oils used in oil-extended rubber tires is 40%, but it may be as high as 55%. Tires may contain from about 40 to 80 parts of carbon black per 100 parts of rubber, 2 to 10 parts per 100 of zinc oxide, 1 to 5 parts per 100 of elemental sulfur, 0.5 to 2 parts per 100 of accelerators, and 0.5 to 2 parts per 100 of antioxidants along with other ingredients. Antioxidants which are often employed include phenyl-β-napthylamine, diphenyl-phenyl-enediamine, diotolyethylenediamine, and *N*-phenyl-*N*′-cyclohexyl-*p*-phenylenediamine. Useful accelerators include 2-mercaptobenzothiazole, *n*-ter-butyl-2-benzothiazole sulfenamide, tetramethylthiuram disulfide, and the zinc salt of 2-mercaptobenzothiazole. Different parts of the tires, for example, the treads, side walls, and inner linings, may contain differing quantities of natural and synthetic rubbers as well as diverse adjuvants in varying amounts. Moreover, different manufacturers may employ different rubbers and different compounding agents. It will be appreciated therefore that "used tires" as a raw materials source represent an exceedingly complex mixture. It is a particular advantage of this invention that it is able to utilize such a complex source.

The actual substrate used in the practice of this invention is a used tire mesh. It is prepared in a known manner from used tire carcasses. Usually, any metal components of the tires (such as the metal bead) are removed, and the remainder is then ground to a grain size of about 1 in. The particles are heated in boiling, saturated calcium chloride to remove the fiber content (rayon, polyester, or nylon threads) and the resulting product recovered, dried, and ground to about 35 mesh. This particulate product will be referred to herein as "used tire mesh". It should be understood, however, that the term is not intended to be limiting; it is only a convenient term used to include starting materials which are principally obtained from used tires but may also be obtained from equivalent sources such as inner tubes and rubber belts.

APPENDIX V

FUEL COMPOSITION

ABSTRACT

A mix of rubber particles and coal gives an improved combustible fuel product for use in coal-burning heating systems. This fuel has a heat value substantially greater than that of coal and exhibits low pollution emissions. The fuel consists essentially of lignite (or brown coal) and cryogenically reclaimed rubber tire fragments. In the preferred range, the fuel consists in weight percent of between 85 and 92% lignite and between 8 and 15% cryogenically reclaimed tire fragments.

U.S. Patent 3,864,094 describes clearly the advantages of using rubber as a fuel. It reads as follows:*

This invention relates to improved fuel products for use in coal-burning heating systems.

At one time coal, both anthracite and bituminous, was the most common fuel for heating and energy conversion systems. For reasons of economy and convenience, oil and natural gas have replaced coal in many fuel applications.

At the present time many coal-burning systems still exist. One problem faced by the operators of these systems is that coal is considered to be a dirty fuel. It can produce dense smoke, particulate matter, and other environmentally undesirable emissions. For these reasons, substitute fuels are being sought which have the same heat value as coal and which are environmentally desirable.

Recently there has been much publicity about the "energy crisis" where the cost of natural gas and oil has been rising and reserves have been diminishing. The reserves of coal are quite large and, in view of the rising gas and oil prices, coal is once again being reviewed as a possible fuel.

It is therefore an object of this invention to provide an inexpensive coal-base fuel which can be used in present coal-burning installations. This fuel has a heat value comparable to that of the coal originally intended for use in the system.

Another object of this invention is to provide an environmentally desirable fuel which will meet the current strict air-pollution emission requirements. Yet another object is to provide a fuel which can be used in existing systems without conversion thereof. These and other objects will become apparent from the following description and appended claims.

The fuel is an admixture of a coal-like material, lignite, and cryogenically reclaimed rubber tire fragments.

Lignite, sometimes referred to as brown coal, is a material which is said to be intermediate peat and bituminous coal and has the texture of the original wood distinct on its surface. It has a heat value lower than that of anthracite or bituminous coal and is believed to have lower pollution emissions. Moreover, it is believed that there are extensive reserves of lignite.

In order to raise the heat value of the fuel products above that of the lignite itself, cryogenically reclaimed rubber tire fragments are mixed with the lignite. The fragments, being in large part carbon, have a higher heat value than the lignite, and there-

* Locketz, H. B., U.S. Patent 3,864,094.

fore may be added in amounts effective to raise the heat value of the mixture to that approximately equivalent to the anthracite or bituminous coal.

Cryogenically reclaimed rubber tire fragments are produced by a process in which rubber tire scraps (i.e., complete tires or parts thereof) are exposed to a cryogenic temperature for a time sufficient to lower the temperature of the scrap to a point at which the rubber in the scraps exhibits brittle fracture characteristics, which is less than about −62°C (−80°F). The embrittled scraps are then transferred to a mill in which the scraps are struck with a force sufficient to shatter them into small fragments. Normally, the tire scraps are exposed to temperatures approximately equivalent to that of liquid nitrogen — i.e., −195.6°C (−320°F) — and then transferred to a rotary hammer mill for fragmentation. About 95% of the fragments produced by this process are less than ¼ in.^{2} and are substantially pure rubber from the tread of the tire or from the tire carcass. The tire cords (fabric, metal, or glass) from the carcass can be separated from the rubber fragments.

The rubber fragments are added to the lignite in amounts sufficient to raise the heat value of the mixture to a level substantially equivalent to the anthracite or bituminous coal. In the preferred embodiments, the fuel product consists essentially, in weight percent, of 85 to 92% lignite and 8 to 15% cryogenically reclaimed rubber fragments. A specific fuel product composition which is suitable for use in most coal-burning installations consists, in weight percent, of 90% lignite and 10% rubber fragments.

This fuel may be used in large coal-burning installations such as electricity-generating plants, which use enormous quantities of coal, and which have been characterized as extreme polluters. This patent claims the following:

1. That this is an improved combustible fuel product, for use in coal-burning heating systems, characterized by a heat value substantially equivalent to coal so as to permit substitution thereof in said heating systems, which burns with low emissions so as to reduce polluting discharges, and consists essentially of lignite and cryogenically reclaimed rubber tire fragments in an amount effective to adjust the heat value of the fuel to substantially the heat value of the substituted-for coal, said fragments being obtained by exposing rubber tire scraps to cryogenic temperatures so as to embrittle the rubber therein and striking said embrittled tire so as to shatter said tire into fragments of a size sufficiently small for use as a fuel.
2. A fuel as in Claim I which consists essentially, in weight percent, of between about 85 and 92% lignite and 8 and 15% cryogenically reclaimed rubber tire fragments.
3. A fuel as in Claim I wherein said cryogenically reclaimed fragments is present in an amount approximately equal to 10%.

APPENDIX VI

ROAD REPAIRS: TIRE ASPHALT RUBBER MIX (B_{11}) BENEFITS

I. INTRODUCTION

The U.S. has entered an era of road maintenance and repairs; the major road construction work of the past decades has, for the most part, been completed. As time passes, these many roads built in the construction era will develop cracks, potholes, unevenness, and other failures.

To extend the useful life of deteriorating roadways, generally accepted restoration typically involves the application of a thin asphaltic overlay over the cracked and otherwise deformed pavement. Historically, however, the application of these thin overlays (generally of 4 in. or less) has resulted in a new complex problem known as "reflective cracking" — defined as the migration of a subsurface cracking pattern into and subsequently through the overlay structure. Once the overlay is fractured, general erosion occurs, which severely affects performance and requires further and costly maintenance.*

The use of waste tire rubber in road-repairing appears to have significant benefits in avoiding the costs, inconveniences, and road hazards involved in the repeated repair of reflective cracks. In a study of various methods of road repairs, Arizona found that asphalt rubber repairs reflected only 4% of the underlying cracks in 3 years, while control sections reflected 17% of the cracks. The Arizona experiences indicate that road repaired with tire asphalt rubber stay repaired for quite a while:

> "The surface is in excellent condition and shows only minor crack reflection after eight years of service."
> "After six years of service, this Project has required no maintenance and shows only a few minor reflective cracks."
> "The asphalt rubber process includes, along with semi-conventional repairs, the placement of ¼-in. overlay consisting of 25% worn tire rubber and 75% asphalt. The engineering details of the process are described in the literature."**

The Arizona projects might have been successful due to the favorable climate of the Phoenix area where they were carried out. Accordingly, in August of 1973 a tire asphalt rubber road repair project was carried out at a severe winter weather location, Flagstaff, Ariz. Flagstaff is at an elevation of 7200 ft with temperatures as low as −5°C (−40°F) and frost depths to 3 ft. The project was reported in 1976 to have "performed excellently with zero maintenance to date".** It appears then, that the tire asphalt rubber road repair process will be valuable in a range of climates.

The highly favorable experiences of Arizona, however, have not yet been duplicated in a significant number of other states. Some tire asphalt rubber road repairs have been made in California and in South Dakota. California has indicated that tire asphalt rubber seems to work as intended. South Dakota has had poor results with the process.

* U.S. Environmental Protection Agency, Scrap Rubber Tire Utilization in Road Dressings, EPA 670/2-74-014, U.S. Environmental Prrotection Agency, EPA Technology Service, March 1974.

** McDonald, C., Prevention of Reflective Cracking in Arizona Minnetanka-East, Rep. No. 11, HPR-1-13(244), Arizona Department of Transportation, May 1979.

These may be due to the use of alternative quantities and types of rubbers used in repairs and/or to different methods of application. A 4-year U.S. Environmental Protection Agency Project to document experience with the tire asphalt rubber repairs is in process and will be completed in 1981.

The Arizona and California experiences provide reasonable documentation with which to estimate the benefits of the tire asphalt rubber road repair process.

II. TIRE ASPHALT RUBBER REPAIRS

The benefits of the tire asphalt rubber repairs might be represented as the road repair costs avoided by use of the process.

Phoenix "streets require a new (conventional) seal coat every 3 to 5 years. The asphalt rubber (tire rubber asphalt) seals have exceeded 7 years to date and it appears that they will last at least 10 years."

According to these estimates, in 10 years we might need one tire asphalt repair or two or three conventional repairs.

The city of Phoenix, in 1972, reported the costs per city block, of conventional repairs, to be $1900. Comparable tire asphalt rubber repairs were reported to cost $2400. Arizona, in 1976, reported: (1) costs of $2.95/yd^2 for a (conventional) 3-in. asphalt concrete overlay with a ½-in. asphalt concrete finishing coat, and (2) costs of $2.32/yd^2 for 1¼ in. of asphalt concrete followed by about ¼ in. of the tire asphalt rubber chip coating and ½ in. asphalt concrete finishing coat. These costs include "the total of all ingredients and operations and are estimations based on . . . a size job . . . generally more than 40,000 square yards." The cost of the tire asphalt rubber chip seal alone accounts for about $0.75 of the $2.32 on the average. California reported, in 1976, average costs of $1.91/yd^2 for 1 to 1½ in. of asphalt overlay. We multiplied 1976 cost figures times the 3733 yd^2 (3121.26 m^2) of the Phoenix block mentioned earlier in order to obtain 1976 "costs per block" useful in estimating the benefits of the tire asphalt rubber process. According to Arizona estimates, a "conventional" repair of the Phoenix block would cost $11,013 and would last as little as 3 years. Based upon the Phoenix experience, we estimate a tire asphalt rubber interlayer repair would cost $8661 and would last 10 years. According to the California information, conventional repairs to a block would cost $7130 and would last as long as 7 years. We regarded the $8661 asphalt rubber cost as representative for California repairs.

III. ARIZONA ROAD REPAIR BENEFIT CALCULATION

Demolition and reconstruction of a city block, in 1977, cost in the neighborhood of $86,000; sometimes reconstruction is necessary. Relatively thin repair overlays are suitable temporary substitutes for reconstruction as long as the coats do not raise the road so high as to cover curbs, storm drains, etc. One 2-in.-thick tire asphalt rubber overlay, according to Arizona experience, can last 10 years. Three 3½-in. thick conventional overlays, a total of 10½ in. thickness might be required in 10 years as an alternative. Assuming that after 10½ in. of road thickness build-up occurs the road must be reconstructed, we can say that five tire asphalt rubber road repairs may be used in place of both nine conventional 3½-in. seals and two major road constructions (Table 1). In addition to the costs avoided, we might say that the highway repair agency could invest these funds not spent to earn money during this period.

We took the present value of the repair costs avoided in 10 years (the 10th to 20th years in Table 1 as if year 10 were the present time) plus the present value of the interest earnable in 10 years on the savings from costs avoided, as one estimate of the total

Table 1
ROAD REPAIR AND RECONSTRUCTION MODEL COSTS

Beginning of year	Conventional repairs costs (3½ in.)	Tire asphalt rubber repairs costs (1¾ in.)	Cost savings
0.00	$86,000 first cycle	$86,000	—
3.33	—	—	—
6.67	—	—	—
10.00	$11,013	$ 8,661	$ 2,352
13.33	$11,013	—	$11,013
16.67	$11,013	—	$11,013
20.00	$86,000	$ 8,661	$77,339
23.33	—	—	—
26.67	—	—	—
30.00	$11,013	$ 8,661	$ 2,352
33.33	$11,013	—	$11,013
36.67	$11,013	—	$11,013
40.00	$86,000	$ 8,661	$77,339
43.44	—	—	—
46.67	—	—	—
50.00	$11,013	$ 8,661	$ 2,352
53.33	$11,013	—	$11,013
56.67	$11,013	—	$11,013
60.00	$86,000 second cycle	$86,000	—

Note: This assumes (1) that original and reconstruction work lasts 10 years, and (2) that a second asphalt rubber repair (and the third, fourth, and fifth) would last 10 years, as does the first. This is not certain.

incremental benefits, in 10 years, of tire asphalt rubber repairs to one city block. This did not include the possible substantial savings in major reconstruction beginning in year 20. We multiplied this by 4/10 to convert this to an average rate per 4 years.

We divided this result by the number of worn tires used in tire asphalt rubber repairs to a city block. The result was an estimate of the benefits of the tire asphalt rubber process per waste tire utilized and per 4 years. We added to this the value of steel recovered from the waste tire in processing.

The present value of the repair costs avoided (cost savings) in 10 years was represented as

$$C_{sp} = (R_c - C_{tra}) + \frac{R_c}{(1 + i)^{3.33}} + \frac{R_c}{(1 + i)^{6.66}}$$

where R_c = conventional repair costs to one city block and C_{tra} = tire asphalt rubber repair costs to one city block.

At a discount rate of 10%, and 1977 costs — R_c = \$11,013 and C_{tra} = \$8661 — the benefits are

$$(\$11{,}013 - \$8661) + \frac{\$11{,}013}{(1.1)^{3.33}} + \frac{\$11{,}013}{(1.1)^{6.66}} = \$17{,}060$$

The present value of the interest earnable on the funds made available by avoiding these costs was represented as

$$S_{ip} = \frac{(R_c - C_{tra})\,(1 + I_g)^{10} - 1}{(1 + I_g)^{10}} + \frac{R_c(1 + I_g)^{6.66} - 1}{(1 + I_g)^{10}} + \frac{R_c(1 + I_g)^{3.33} - 1}{(1 + I_g)^{10}}$$

$$S_{ip} = \frac{(\$11{,}013 - \$8661)\,(1.1)^{10} - 1}{(1.1)^{10}} + \frac{\$11{,}013(1.1)^{6.66} - 1}{(1 + I_g)^{10}} + \frac{\$11{,}013(1.1)^{3.33} - 1}{(1.1)^{10}}$$

$$S_{ip} = \$5{,}739.20$$

The combined sum of the costs savings and interest benefits is \$22,799.63; the benefits each 4 years, including the revenues from recovered steel, per solid waste tire are

$$\text{Benefits} = \frac{\$22{,}799.63}{N_b}\left(\frac{4}{10}\right) + \$0.07 = \$37.44$$

where N_b = the number of solid waste passenger car tires used in tire asphalt rubber repairs to a city block, $N_b = W_p P G A_b N$, N_b = pounds tire rubber/gal (gal/yd²) (yd²/one city block) = number of worn tires/one city block, and W_p = the proportion, by weight, of tire asphalt rubber which is worn tire rubber; P = the weight, in pounds per gallon, of tire asphalt rubber; G = the application rate, in gallons per square yard, for asphalt rubber; A_b = the area, in square yards, of a city block (Phoenix, Ariz., 3733 yd²); and N = the proportion of a tire that it takes to recover 1 lb of tire asphalt rubber.

An example calculation follows:

$$N_b = (0.25)\left(\frac{7.5\ \text{lb}}{1\ \text{gal}}\right)\left(\frac{0.5\ \text{gal}}{1\ \text{yd}^2}\right)\left(\frac{3733\ \text{yd}^2}{\text{one city block}}\right)\left(\frac{0.07\ \text{worn tires}}{\text{pounds crumb rubber}}\right) = \frac{244\ \text{tires}}{\text{one city block}}$$

These gross benefits are substantial. They are, however, misleading with respect to *current* tire asphalt rubber road repairs since the rubber currently used is but *tread* rubber ground from worn tires. In current tire asphalt rubber practices, many worn tires, less some tread rubber, remain to be disposed of. Current tire asphalt rubber repairs use 60 lb bags of tread rubber which were ground from 175 worn tires. The proportion, "n", above, for this procedure is 2.917; the number of tires used in a city block at this rate is 10,208; the gross benefit per tire drops to \$0.89 per tire in this case, and in addition, 8,165 tires per city block repaired still remain to be disposed of. Without utilizing the entire amount of rubber available in the worn tire, the benefits of tire asphalt rubber, on a per-tire basis, are much lower.

Cryogenics can be used to separate worn tires into three parts — rubber, metal, and fabrics — so that virtually all of the tire rubber can be recovered. Steel-belted radial tires are processed by cryogenics just as easily as nonsteel-belted tires. Cryogenics, together with solid separation systems, produces salable metals and fabrics in addition: 3 lb of steel and 3 lb of fiber may be recovered for each waste tire processed. These provide additional gross benefits for the cryogenics/road repair alternative. Of course, the costs of cyrogenic processing must be included in the analysis in the appropriate place.

The 3 lb (0.0015 tons) of steel recovered is worth \$0.07 when valued at current prices of \$45/ton; no value data for the fibers recovered was available. The benefits of a tire asphalt rubber process, then, are B_{11} = \$37.44 per tire per 4 years. This includes \$37.37 costs avoided and interest savings plus \$0.07 for recovered product values.

Each worn tire processed for use in tire rubber asphalt eliminates one *waste* tire and its associated processing costs. Waste processing costs are \$0.92 for a tire landfill. Consequently, the tire rubber asphalt alternative realizes an additional \$0.92 benefit per tire each 4 years.

The total road repair and decreased waste benefits for the road repair tire handling alternative, B_{11} = \$38.36.

IV. CALIFORNIA ROAD REPAIR BENEFIT CALCULATION

The asphalt rubber benefits are smaller according to California road repair practices and/or when treated in a more conservative fashion. Going on the assumption that construction or reconstruction lasts 10 years, conventional repairs last 7 years, and asphalt rubber lasts 10 years, Table 2 was prepared.

This table probably illustrates the current perception of most highway repair officials with respect to the asphalt rubber process. It appears in any given year to be more expensive (\$8661 vs. \$6366) and does not provide any benefit in terms of decreased costs. The present value (10%) of the costs for the two cases are approximately equal.

The 1½-in. asphalt concrete and 1¾-in. asphalt rubber repairs are not comparable, however. The asphalt rubber roads, as indicated above, have fewer reflected cracks. In addition, the asphalt rubber repairs are fewer with less nuisance and accident hazards created, and, as indicated above, it may be that asphalt rubber lasts more than 10 years. Consequently, we used the Arizona estimates in calculating the benefits of the asphalt rubber process. We investigated a lower conventional repair process cost (R_c = 6366) in both the framework of Table 2 and the formulas given.

V. LIMITS OF THE TIRE RUBBER ASPHALT PROCESS

The number of worn tires which may, potentially, be used in tire asphalt rubber repairs each year in the U.S. may be symbolically represented as:

$$T_w = 587 R_L R_e N_s{}^*$$

where R_L = the mileage of cracked roads repaired, temporarily, each year (as opposed to rebuilt roads), R_w = the average width, in feet, of a U.S. road, and N_s = the number of worn tires used in 1 yd² of tire asphalt rubber road repairs. N_b/A_b

Data on the mileage of temporary repairs carried out each year in the U.S. are not readily available. It is possible to gain insight on the market for the tire rubber asphalt repairs by a calculation of the number of miles which could be repaired with the 200 million waste tires generated each year. To do this we set "T" equal to 200,000,000, assumed an average road width of 40 ft, and solved for R_L:

$$R_L = \frac{T_W}{587\ R_W N_S}$$

$$R_L = \frac{200{,}000{,}000}{587(40)(0.065)} = 131{,}044 \text{ miles}$$

Each year 131,000 mi (210,821 km) of road repair work will be needed to absorb all of the solid waste tires generated. U.S. streets and roads, in 1974, accounted for 3,815,807 mi (6,139,633 km); almost all (3,000,000 mi or 4,827,000 km) of these roads are asphalt. All of the worn tires currently generated could be used in the tire asphalt rubber process, assuming that a road requires repairs each 3 years, and that 15% of

* The constant 587 is used to convert the term $R_L R_w$ to square yards.

Table 2
CALIFORNIA ROAD REPAIR COSTS

Year	Conventional (1½ in. AC)	Tire asphalt rubber (1 in. AC + ¼ in AR + ½ in: ACFC)
0	86,000	86,000
10	6,366	8,661
17	6,366	
20		8,661
24	6,366	
30		8,661
31	6,366	
38	6,366	
40		8,661
45	6,366	
50		8,661
52	6,366	
59	86,000	86,000

Note: AC, asphalt concrete; ACFC, asphalt concrete finishing coat; AR, asphalt rubber.

these repairs are needed for fatigue-type cracking of the sort controlled by asphalt rubber. If the average time between road repairs were 5 years, 90,000 mi (144,810 km)/year would need repairs; for an average life of 7 years, 57,500 mi (92,517 km)/year would need repairs. It would appear, then, that tire asphalt rubber is potentially a fairly large-scale process.

APPENDIX VII

REMOVAL OF BUILT-UP LAYERS OF ORGANIC COATINGS*

ABSTRACT

This invention relates to the removal of layers of organic coatings built up upon the surfaces of carriers or supports, and relates more particularly to the recovery of carriers or supports by the removal of undesirable deposits which progressively built up on the surfaces thereof.

In many industrial processes it is desirable to apply one or more surface finishes or surface treatments to parts. Such surface finishes, such as base coatings or finished color coatings, and many surface-treating compositions contain organic binder components to effect adherence of such coatings or finishing compositions to the surfaces of the parts being coated or treated. When the parts are being coated or treated by a spray process or electrostatic coating process, the part is usually supported on a pedestal during exposure to the spray coating or electrostatic coating operation. When the parts are coated or treated by a dip coat, flow coat, spray coat, or electrostatic coating process, it is the usual process to suspend such parts from a hanger rod during exposure to such coating or treating operations.

It is impractical to mask the parts in a manner to restrict the coating or treating compositions to the portions being coated or treated. As a result, some portion of the coating composition or treating composition finds its way onto the support or hanger rod by which the article is supported during exposure to the coating or treating material.

Such support or suspending device is used repeatedly for such operation with the result that the coating or treating composition builds up on the support or hanger rod. Oftentimes the support rod continues to support the article during a subsequent processing beyond the coating or treating step, such as through a baking step wherein the coating article is subjected to a drying and heating cycle for setting or cooling the applied coating. Such heating step serves to bake and rebake the coating or treating materials that build up on the article support.

Soon the material built up on the support renders the support incapable of use in the absence of the removal of the cake of material built up on the surfaces above. The removal is rendered difficult by reason of the presence of the organic binder which operates to interbond the built-up layers one onto the other and to the surfaces of the support. Removal is rendered even more difficult where the organic binder is heat-sensitive and the supports have been subjected to repeated heating cycles during support of the coated or treated article through a baking or cooling operation.

The problem arises most frequently in continuous operations for color coating to produce finished parts wherein use is made of an endless chain from which hanger rods are suspended in longitudinally spaced relation. The chain winds its way through a mounting station wherein the articles are hooked onto the rods, a coating station wherein the coating composition is applied (as by spray coating, dip coating, flow coating, electrostatic coating, etc.), a drying zone wherein the coating is allowed to dry in whole or in part, a baking zone wherein the coated articles are carried through a drying or baking oven maintained at elevated temperature, and an unloading station

* Braton, N. R. and Acker, J., U.S. Patent 3, 934, 379, Removal of Built-up Layers of Organic Coatings.

wherein the coated articles are removed from the hanger rod and returned to the loading zone where the travel through the cycle of operations is repeated.

The problem of removal is further complicated by the fact that, from time to time, the composite built up on the support may differ from layer to layer through the cross section by reason of the use of the support for articles treated or color-coated with compositions that differ one from the other.

Current practices for removal of such built-up composites from the surfaces of the supports is by emerging supports in a liquid medium such as a solvent, to dissolve the material off the support.

Such techniques face a number of objections: the solvents or chemicals are not only relatively expensive but considerable care must be exercised in their handling and use because of their combustibility, the corrosive nature of the chemicals, and the fumes that are given off. Considerable amount of equipment and energy is required and oftentimes long periods of soaking are necessary to effect sufficient alteration of the materials of the composite to enable removal. This not only requires the expenditure of considerable labor for handling, but the residual liquids and solids present serious problems of disposal. In general, the entire recovery procedure is unattractive from the standpoint of the messy materials that are left for disposal.

Another technique that has been tried makes use of oxidative combustion of the organic matter by heating the entire assembly in a combustion furnace in which an oxidizing atmosphere is maintained. This technique is limited in its practice to supports formed of noncombustible material such as steel or iron. A considerable amount of heat is required to raise the temperature of the entire assembly, including the metal support as well as the composite to combustion temperature, thereby materially increasing the cost of recovery.

Such combustion procedures are relatively slow, and considerable amount of time is consumed in loading and unloading the furnace and heating up to combustion temperature, and special equipment is required for such purposes. Aside from the above, the supports are sometimes undesirably altered by the temperature conditions at which they are exposed and oftentimes an undesirable residue is left on the support which becomes even more difficult to remove.

Thus it is an object of this invention to provide a simple and efficient method for the removal of organic deposits built up on the type of supports described. The deposits can be cleanly removed from the supports in a manner which raises no problem of material handling or disposal of the composite material following separation of the support. This is accomplished without special handling and with a minimum amount of labor or additional space or equipment. The entire operation can be carried out without the need for removal of the supports from their normal arrangement when in use and adapted to a continuous operation as an adjunct to the in-line operation of the supports during their normal use in coating or treating the surfaces of articles. The removal of the built-up composites is rapid and complete thereby leaving a clean support for subsequent reuse, and in which the process is capable of being practiced with wooden or similar supports, fabricated of natural or synthetic materials as well as material supports, thereby enabling multiple use of the process of this invention.

Briefly described, the concepts of this invention reside in the lessening of the bonded relationship at the interface between the composite and surface of the support whereby clean separation of one from the other can be effected. This, coupled preferably with the embrittlement of the composite interface, enables removal of the organic composite to be effected in response to simple impact or abrasion or a combination of both for a separation of the composite as a solid material.

The lessening of the bonding relationship between the composite of the organic material and the surfaces of support relies on their differences in coefficients of thermal expansion and contraction whereby a sudden change in temperature of one or the other or both will bring about differences in the amount of expansion or contraction thereby straining the bonded relationship between the composite and the support at their interface.

In accordance with the preferred practice of this invention, the desired temperature differential is effected by a sudden reduction in the temperature of the support or the composite or both, but it is preferred to effect rapid cooldown of the material making up the composite to effect embrittlement of the composite simultaneously with the lessening of the bonding relationship between the composite and the support at their interface.

As the coolant for rapid reduction in temperature, use should be made of a liquid, preferably a liquefied gas having a boiling point below −51.11°C (−60°F) at atmospheric pressure and preferably liquefied gas, such as liquefied air, liquefied nitrogen, or carbon dioxide. Each of these is available in unlimited amounts and at a relatively low cost, some as for a by-product for recycling gaseous materials such as liquid oxygen which finds major use in the oxygen converter process used in the manufacture of steel. The gaseous products that are given off in response to the utilization of such liquefied gas as coolants are inert gases which can be freely fielded off into the atmosphere.

The desired rapid temperature change can be effected in a number of ways. When the support is formed of a material which is highly heat-conducted, the support can be immersed in whole or in part in a bath of the liquefied gas. For example, portions of the rod or support can be dipped into the bath of liquified gas whereby the end portion will be rapidly reduced in temperature. Such reduction in temperature transfers by conduction through the rod or support to the portions underlying the composite coatings to effect reduction in the bonded relationship and at least partial embrittlement. Such whole or partial immersion of the support in the bath of liquefied gas can be done piece by piece, or the entire process can be carried out as a continuous operation by providing a bath in a line having a drop portion for dipping the supports into the bath.

It is preferred to effect the desired temperature change by application of the liquefied gas directly onto the composite coating. This can be achieved by a flow-coat process or by a dip-coat process, but it is preferred to effect application by a spray process wherein the liquefied gas is sprayed directly onto the surface of the composite coating to effect rapid refrigeration thereof as the liquefied gas evaporates from the wetted surfaces. This not only effects the desired embrittlement of the material making up the composite layer, but a saving is experienced in the amount of liquefied gas that is used since the cooling down of the metal base or support is not required. Such application by spray-coating does not require a facility to collect the overflow or excess since the amount sprayed onto the surface can be controlled to that which is fully converted to the gaseous state and for effecting the desired amount of refrigeration.

Application by spray-coating or flow-coating can be effected while the support is suspended or otherwise supported on a base. However, the process of spray- or flow-coating the surfaces of the composite layers built up on the supports enables application of the liquefied gas without removal of the supports from their normal arrangement and position of use on the endless chain or conveyor. The supports or hanger rods can be sprayed or otherwise wetted with the liquefied gas as they advance through an application zone to effect the rapid refrigeration of the composite layer. The amount of time for refrigeration will depend upon the temperature of the liquefied gas, the manner of application, as well as the thickness of the composite layer. For layers greater than ¼ in. thickness, the application of liquid nitrogen for 24 sec would be sufficient to effect the desired rapid refrigeration.

Removal of the composite layer, after refrigeration, can also be carried out in a number of ways. In the event that removal is being made of composite material from the surfaces of the hanger rod or support, clean removal can be effected by means of a blast of abrasive material directed onto the composite layer immediately following refrigeration. Such abrasive particles can be blasted onto the surfaces by means of an air blast, using a conventional air gun, or by means of an airless blast using a centrifugal wheel by which means abrasive particles such as sand or metal shot are drawn radially outwardly at a at a high speed from radially extended blades mounted in a rotating wheel. Such airless centrifugal blasting means are well known to those skilled in the art such as Wheelabrators® manufactured by Wheelabrator-Frye Inc. of Mishawaka, Ind.

In the event that the composite layer is of a thickness too great for efficient removal by abrasive particles blasted onto the surface, removal can be effected by impact after refrigeration by an amount to cause embrittlement of the organic materials making up the composite layers. Such impact can be effected by means of a hammer or other tool which, because of the brittle character of the composite, will shock the material to cause breakup into small pieces which fall from the support. Impact, such as by means of a hammer, can be applied to the rodder support or directly onto the composite layers if necessary. Often a combination of impacts will be sufficient to remove the built-up layers and substantially clean the hanger rod support to its original condition.

The blasting or impacting, to effect removal, can be carried out as a continuous operation with the previous step of refrigeration, or it can be carried out as an intermittent operation on each piece as it is refrigerated. In any event, such impacting or blasting should be effected immediately following refrigeration or during the last stages of refrigeration so that the composite material will be in the refrigerated state during such removal operation. The removed composite material falls gravitationally from support and can be collected in suitable pans or hoppers for dispersal. With thin layers, it is sufficient to freeze to effect unbonding of the layer to removal by blasting; with thicker layers, it is preferred to freeze to effect unbonding and to embrittle to removal by impact.

APPENDIX VIII

THE ECONOMICS OF LIQUID NITROGEN FREEZING*

I. INTRODUCTION

Linde** Liquid Nitrogen Food Freezing System will produce a better-quality product at a lower overall cost. Better quality is an established fact and can be proven on your own product. A comparison with mechanical air blast refrigeration can define the savings.

All the costs of cryogenic freezing are easily identified. Unfortunately, many of the true costs of mechanical refrigeration are too often overlooked or forgotten. The following analysis will indicate most of them.

II. CAPITAL INVESTMENT

The initial capital outlay required for cryogenic freezing equipment is far less than the cost of a comparable mechanical system. The dollar savings are important to you. Interest payments are reduced if the equipment is financed, or cash outlay is lowered if the equipment is purchased. In either case, additional capital is available for your discretionary use.

Linde Food Freezing Systems can be purchased with only the capacity needed for immediate production goals. Additional capacity is quickly and inexpensively added as your markets expand, thereby keeping costs in line with expansion.

Unfortunately, due to the large size of the components, great expense, and length of delivery time required to increase the capacity of a mechanical air blast freezer, it becomes a major task to grow with your market.

III. DEHYDRATION LOSS AND SHRINKAGE

All products lose some weight during the freezing process due to drip and evaporation of water vapor into the surroundings. The amount lost varies appreciably with the type of freezing process used. Loss will be greatest with typical mechanical air blast freezers because of the relatively long time during which the warm food surface is exposed to the dry surrounding air. Typical weight losses of 3% or more are experienced. This means that for 100 lb of meat patties processed, 97 lb or less are available for sale. Lost weight is lost profit.

On the other hand, the use of a Linde Liquid Nitrogen Freezing System lowers product surface temperature rapidly, and the total freezing process takes only a few minutes. This minimizes shrink and reduces weight loss, usually to ½% or less. Equally important, dehydration has not compromised taste or texture.

The savings realized from lack of dehydration loss alone could pay for a large part of the operation of a Linde Liquid Nitrogen System, depending on product value and volume.

* Food Technology, 10/77-0756-5m; F-4177 and Cryogenic Batch Chiller Systems, F3962-6176-3m-86-0414, Union Carbide Corporation, New York, N.Y.

** The term Linde is a registered trademark of Union Carbide Corporation.

IV. LABOR SAVINGS

Mechanical refrigeration requires a large freezer compartment, compressors, motors, cooling towers, and electrical equipment. Since a Linde Liquid Nitrogen system requires none of these, substantial labor savings can be realized in several areas.

A Linde Liquid Nitrogen System requires minimal operator attention and runs virtually unattended. Also, a liquid nitrogen tunnel is ideal for a completely automated "in-line" production system from prefreezer handling to the holding freezer.

Maintenance of a Liquid Nitrogen Freezer is minimal. Mechanical systems require many man-hours of maintenance each year, as well as a considerable spare-parts inventory and shop facilities.

Cleaning of a Liquid Nitrogen Freezer can be done in less than one man-hour by unskilled labor. The freezer can be placed back in service 15 min after start-up. A mechanical system may require three men for a full 8-hr shift and then an additional 8-hr shift for cooldown. Consider not only the labor costs but the downtime.

It is estimated that, all factors considered, a direct labor savings of at least 50% will be realized using cryogenic freezing.

V. FLEXIBILITY

With Linde Liquid Nitrogen, production scheduling can be planned around supply and marketing conditions, not refrigeration capacity. Production rate is elastic and can grow as business grows. Costs can be related directly to sales and become controllable operating expenses rather than fixed overhead.

VI. ELECTRIC POWER

Mechanical freezers run 24 hr a day whether your product is being processed or not. With a Linde system, cryogenic gases are used to freeze the product only during actual freezing operations. You have complete control of freezing costs and an insignificant electric bill.

VII. FLOOR SPACE

The bulky freezer rooms, compressors, motors, and cooling towers of mechanical freezers require at least five to seven times as much floor space as that required for a cryogenic freezer. Although often overlooked, this also can mean dollars. Valuable floor space is wasted, both indoors and out, which could be used for other operations or plant expansion.

APPENDIX IX

LIQUID NITROGEN TOMATO PEELING*

I. INTRODUCTION

Tomatoes in which the skin has been "crust-frozen" can be quickly, completely, and easily peeled after they have rapidly thawed. The development and use of liquid nitrogen to crust-freeze the tomato can be a viable and economical process with benefits as described in the following sections.

II. CHANGES MADE POSSIBLE BECAUSE OF CRYOGENICS

The process of peeling tomatoes by using low temperature-short duration freezing has been under development for considerable time. Liquid nitrogen can be successfully used as the refrigerant for freezing the skin and a thin layer of cells just beneath the skin. Upon being immediately thawed, the tomato skin can be quickly and easily removed. "Liquid nitrogen-peeled" samples can be evaluated on the basis of loss of peel and trim, and these losses can be compared with samples peeled in boiling water. The peel and trim losses from liquid nitrogen-peeled tomatoes are reduced by approximately 50%. There is a decrease in the percentage of broken fruit in canned tomatoes which have been "liquid nitrogen-peeled." Titratable acidity is higher, and pH is lower in liquid nitrogen samples. Liquid nitrogen-peeled tomatoes exhibited better visual color, although the color of the homogenized canned product is not statistically different. Utilization of calcium chloride to firm the product is more efficient in the liquid nitrogen-peeled tomatoes.

Raw tomatoes are fed into the washer where high-pressure sprays of water clean the product. Transfer by mechanical means is made to the scalder, where the tomatoes are prepared for peeling. Peeling is generally accomplished by using a rubbing action on the surface of the tomato. The product is then packed into a can, and juice and salt are added. The cans are sealed, and the product is cooked, cooled, and placed in a case for shipment.

III. ADVANTAGES AND DISADVANTAGES

Fast, economical removal of peel before canning is a problem in the tomato-processing industry. This problem has been accentuated by recent development of tough-skinned varieties of tomatoes suitable for mechanical harvesting. Methods involving the removal of skin from tomatoes by hand-peeling with knives, the use of boiling water or steam, boiling sodium hydroxide and boiling calcium chloride as treatments prior to peeling, have been used. Currently, virtually all canning operations use a caustic (sodium hydroxide) treatment in the peeling operation. Presently, and in the near future, disposal of this spent caustic solution is, and will be, a very serious pollution problem.

Thermal processes for treatment of tomatoes prior to peeling tend to reduce the quality of the product due to the excessive heat penetration inherent in these processes. Excessive heat penetration not only causes a loss of color but also affects the firmness

* Liquid Carbonic Corporation, Chicago.

and wholeness which must be maintained to obtain a high-quality canned product. The toughness of the tomato skin and the adhesion of the skin to the flesh of the fruit have contributed to the difficulties in obtaining a satisfactory removal of the skin from tomatoes.

Experimental work and development studies have been done utilizing liquid nitrogen as a treatment prior to peeling tomatoes. Tomatoes were allowed to ripen on the vine and were harvested on the day of the test. Selected samples were used for testing so that good fruit was used. Samples were immersed in liquid nitrogen for 20 sec, removed, and immediately thawed 30 sec in room temperature water. Treated in this way, the tomato fruit separated completely from the skin sack, allowing the sack to be slit and easily removed. The peel and trim were removed from the tomatoes and weighed to determine peel and trim loss. Studies have also been conducted which substantiate the applicability of a liquid nitrogen production freezing tunnel to perform the freezing operation. Exact figures regarding liquid nitrogen utilization are not available at this time. However, an estimate indicates that the crusting could require approximately 0.17 lb LN_2/lb of tomatoes. Peel and trim losses were reduced by approximately one half when compared with boiling-water treated samples. The tomato fruit was held loosely in a skin sack in those samples treated with liquid nitrogen, whereas the skin tended to adhere to the flesh of tomatoes treated with boiling water. There was also no tearing or shrinking of the tomato flesh. Since the process was used to freeze the skin and a thin layer of cells beneath the skin, the peel and trim loss was only 6.8% as compared to 12.7% for boiling-water peeled tomatoes.

In a commercial operation, the percent loss in peel and trim in tomatoes peeled by boiling water or other thermal processes would probably be much higher, as overripe, soft tomatoes are usually disintegrated and lost. The liquid nitrogen process for peeling does not subject the tomatoes to the rigorous treatment of heating, and could result in an overall increase in recovery, because of a smaller loss of overripe, soft fruit.

The percent of broken fruit was significantly higher in the samples processed in boiling water and peeled, regardless of the postpeeling treatment. The percentage of broken fruit ranged up to 10% for the water peeled, without additives, whereas no trace of broken fruit was found in the nitrogen-peeled samples. The difference in the amount of broken fruit in the water-peeled samples when compared with nitrogen-peeled samples may be attributed in part to the texture and physical characteristics at the time of canning. The nitrogen-peeled tomatoes possessed a smoother texture, maintained their shape better, and were firmer than the water-peeled tomatoes. Therefore, more damage appeared to have occurred in the water-peeled fruit by physical handling during the canning process. Calcium chloride is routinely added to canned tomatoes to firm the peeled product. Liquid nitrogen-peeled tomatoes possess a higher degree of firmness than boiling water-treated samples, which indicates a greater utilization of the calcium chloride usage.

Firmness and calcium content of canned tomato samples peeled with liquid nitrogen and canned without additives, with sodium chloride and calcium chloride added, was determined, and compared with results obtained from boiling water-peeled samples treated the same way. The liquid nitrogen-peeled samples had a greater shear press value (firmness) and maintained their shape better than comparable samples peeled with boiling water.

The samples of tomatoes with calcium chloride added were firmer than tomatoes canned plain or with sodium chloride added. However, it must be noted that tomatoes peeled with liquid nitrogen with calcium chloride added were much firmer than comparable samples peeled with boiling water with calcium added. It is assumed that heat destroyed some of the ability of the outside layers of tissue to be firmed by the calcium

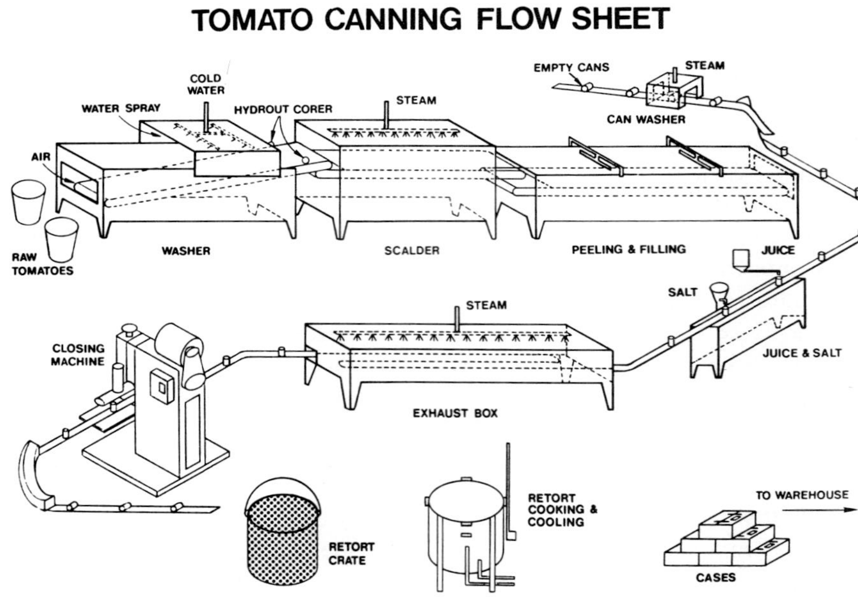

FIGURE 1. Flow diagram of a typical tomato canning operation.

addition. The amounts of calcium present in the canned samples were not, however, significantly different.

Liquid nitrogen-peeled tomatoes are not exposed to the high temperatures that other techniques utilize. Thus no cooking of the product occurs, and better visual color and taste are achieved.

There was no statistical difference found in color between the blended tomato samples peeled with liquid nitrogen when compared to those peeled with boiling water. Although the tomatoes peeled with LN_2 had a redder visual color, the instrument used for measurement did not detect any difference in color.

By using liquid nitrogen for freezing additional benefits could be attained, one of which is the elimination of hot caustic solutions, which would also eliminate, or greatly reduce, disposal problems. Also, the peel from the tomato would be available for use as animal feed or as a soil conditioner.

INDEX

D

E

F

G

H

I

J

K

L

M

N

O

P

Q

R

S

T

U

V

W